FURTHER PRAISE FOR *THE NATURE FIX*

"[Williams] makes a compelling argument for time outdoors. She takes a refreshing approach, including 'forest bathing' (the Japanese custom of a sensory walk in the woods); ecotherapy in Scotland; and how nature can produce the same effects as mind-altering drugs. Thought-provoking and excellent." —*BBC Wildlife*

"Chock-full of insights about the health benefits of spending time in nature. (It turns out that lying on the beach is good for you.)"
—Rohan Silva, *Guardian*, "Best Holiday Reads 2017"

"Williams is an elegant yet witty writer, and she makes a terrific guide."
—*Success* magazine

"A veteran journalist, Williams flexes her conditioned reporting muscles in *The Nature Fix*. Her fast-paced book takes readers across three continents as she trails all kinds of experts, from psychologists to foresters."
—*Earth Island Journal*

"A thoughtful, refreshing book with a simple but powerful message."
—*Kirkus Reviews*, starred review

"[A] powerful environmental call to arms." —*Publishers Weekly*

"*The Nature Fix* is a beautifully written, thoroughly enjoyable exposition of a major principle of human life now supported by evidence in biology, psychology, and medicine."
—Edward O. Wilson, University Research Professor Emeritus, Harvard University

"I'm no tree hugger, but *The Nature Fix* made me want to run outside and embrace the nearest oak. Not for the tree's sake but mine. Florence Williams makes a compelling, and elegant, case that nature is not only beautiful but also good for us. If Thoreau were steeped in modern neuroscience and possessed an endearingly self-deprecating sense of humor, the result would be the book you hold in your hands."   —Eric Weiner, *New York Times* best-selling author of *The Geography of Genius*

"Florence Williams, keen observer, deft writer, creates a fascinating mosaic here. What are the costs—to us!—of humanity's increasing disconnection from nature? What are the likely benefits—to us!—of retaining that threatened connection? . . . Large."
—David Quammen, *New York Times* best-selling author of *Spillover*

"[*The Nature Fix*] has much to offer to anyone interested in human health, from city planners and health practitioners to educators, scientists . . . and anyone concerned with their own well-being."
—Esther Jackson, public services librarian at the New York Botanic Garden

THE

Nature
Fix

ALSO BY FLORENCE WILLIAMS

Breasts: A Natural and Unnatural History

THE
Nature
Fix

WHY NATURE MAKES US HAPPIER, HEALTHIER, AND MORE CREATIVE

FLORENCE WILLIAMS

W. W. NORTON & COMPANY

Independent Publishers Since 1923

New York | London

For information about permission to reproduce selections from this book,
write to Permissions, W. W. Norton & Company, Inc.,
500 Fifth Avenue, New York, NY 10110

For information about special discounts for bulk purchases, please contact
W. W. Norton Special Sales at specialsales@wwnorton.com or 800-233-4830

Manufacturing by Quad Graphics Fairfield
Book design by Chris Welch
Production manager: Julia Druskin

Library of Congress Cataloging-in-Publication Data

Names: Williams, Florence, 1967– author.
Title: The nature fix : why nature makes us happier, healthier, and more creative /
Florence Williams.
Description: First edition. | New York : W.W. Norton & Company, Independent
Publishers Since 1923, [2017] | Includes bibliographical references.
Identifiers: LCCN 2016040709 | ISBN 9780393242713 (hardcover)
Subjects: LCSH: Nature—Psychological aspects. | Environmental psychology. |
Creative ability.
Classification: LCC BF353.5.N37 W55 2017 | DDC 155.9—dc23
LC record available at https://lccn.loc.gov/2016040709

ISBN 978-0-393-35557-4

W. W. Norton & Company, Inc.
500 Fifth Avenue, New York, N.Y. 10110
www.wwnorton.com

W. W. Norton & Company Ltd.
15 Carlisle Street, London W1D 3BS

1 2 3 4 5 6 7 8 9 0

To my father, John Skelton Williams, for showing me the

natural world in the first place. You always

made it magical.

CONTENTS

THE
Nature
Fix

The Cordial Air

May your trails be crooked, winding, lonesome, dangerous,
leading to the most amazing view.
—EDWARD ABBEY

I was hiking in Arches National Park when the Mappiness app in my phone pinged me. Some people would be annoyed, but not I. Finally, I was somewhere outside and beautiful and could tell the app how happy, relaxed and alert I was. Very, very and very. I told it so by tapping on the screen. Then I victoriously took a photo of the smooth, salmon-colored cliffs in front of me. Small topographies of lichen poked through a crack. A few perfect white clouds pottered across a French blue sky. Let Big Brother, toiling away in some windowless university lab, eat that for lunch. After many months and 234 interactions with this app, I almost always got pinged when I was indoors and working, which didn't seem very helpful to either the Mappiness project or to my own. (And it didn't seem fair, because I was outside fairly often, wasn't I?) Mappiness is in the midst of a multiyear big-data grab, asking tens of thousands of volunteers to record their moods and activities twice a day at random times. Then it matches those responses to an exact GPS location from which it extracts information on the weather, amount of daylight and other environmental characteristics. The aim is simple: What makes people happy? Does place matter, or not so much?

Big Brother—or Big Scientist, really—is George MacKerron, a young and congenial economist at the University of Sussex. As he explained it to me, much of the happiness data out there involves relationships, activities and economic behaviors, and much of it is familiar: people are happiest when they are well enmeshed in community and friendships, have their basic survival needs met, and keep their minds stimulated and engaged, often in the service of some sort of cause larger than themselves. But MacKerron wondered about the people who already have these things going for them, or, for that matter, about the people who don't; are there other factors that could make meaningful differences in the march of their days?

To find out, he launched Mappiness in 2010 and within a year had gathered 20,000 participants and over a million data points (by the time I joined a few years later, he was up to 3 million). Here's what the data shows: People are least happy at work or while sick in bed, and most happy when they're with friends or lovers. Their moods often reflect the weather (most live in the UK, so that's not surprising). But one of the biggest variables, the surprising one, is not who you're with or what you're doing (at least for this iPhone-using crowd, which tends to be young, employed and educated). It's where you are. As one of MacKerron's papers concludes: "On average, study participants are significantly and substantially happier outdoors in all green or natural habitat types than they are in urban environments." (And, in case you're wondering, the data didn't just reflect a vacation effect, since he factored that in.)

The difference in joy respondents felt in urban versus natural settings (especially coastal environments) was greater than the difference they experienced from being alone versus being with friends, and about the same as doing favored activities like singing and sports versus not doing those things. Yet, remarkably, the respondents, like me, were rarely caught outside. Ninety-three percent of the time, they were either indoors or in vehicles. And even the app's definition of "outside" could mean standing at an intersection or collecting the mail. My own personal data was pretty pathetic. The app caught me exercising or relaxing outside only 17 times, or 7 percent of the pings over the course of a year. Most often I was working, followed by number two, doing childcare, followed by commuting, doing housework and eating (well, at least something was fun). In the midst of a flirtation with meditating, I was caught doing that exactly twice.

What Mappiness reveals—our epidemic dislocation from the outdoors—is an indictment not only of the structures and habits of modern society, but of our self-understanding. As the writer Annie Dillard once said, how we spend our days is how we spend our lives. Why don't we do more of what makes our brains happy? Are we just too knackered by life's demands, too far away from greenery or too tempted by indoor delights, especially the ones that plug in? Partly, but not entirely. In a revealing set of studies at Trent University in Ontario, psychologist Elizabeth Nisbet sent 150 students either outside to walk on a nearby path along a canal, or underground to walk through the well-used tunnels connecting buildings on campus. Before they left, she asked them to predict how happy they thought they'd feel on their walks. Afterward, they filled out questionnaires to gauge their well-being. The students consistently overestimated how much they'd enjoy the tunnels and underestimated how good they'd feel outside. Social scientists call these bad predictions "forecasting errors." Unfortunately, they play a big role in how people make decisions about how to spend their time. As Nisbet rather dejectedly concluded, "People may avoid nearby nature because a chronic disconnection from nature causes them to underestimate its hedonic benefits."

So we do things we crave that make us tetchy, like check our phones 1,500 times a week (no exaggeration, but I will point out that iPhone users spend 26 more minutes per day on their phone than Android users, which may be a good reason to marry an Android user), while often neglecting to do the things that bring us joy. Yes, we're busy. We've got responsibilities. But beyond that, we're experiencing a mass generational amnesia enabled by urbanization and digital creep. American and British children today spend half as much time outdoors as their parents did. Instead, they spend up to seven hours a day on screens, not including time in school.

We don't experience natural environments enough to realize how restored they can make us feel, nor are we aware that studies also show they make us healthier, more creative, more empathetic and more apt to engage with the world and with each other. Nature, it turns out, is good for civilization.

This book explores the science behind what poets and philosophers have known for eons: place matters. Aristotle believed walks in the open air clarified the mind. Darwin, Tesla and Einstein walked in gardens and groves to help them think. Teddy Roosevelt, one of the most hyperproductive presidents of all time, would escape for months to the open country. On some level they all fought a tendency to be "tired, nerve-shaken, over-civilized people" as hiker-philosopher John Muir put it in 1901. Walt Whitman warned of the city's "pestiferous little gratifications" in the absence of nature. Park builder and public-health advocate Frederick Law Olmsted understood. He changed the torso of my hometown and that of many other cities as well.

The Romantic movement was built upon the idea of nature as the salvation of the mortal soul and the mortal imagination, with poets penning odes to high peaks just as industrialization was beginning to choke its way through Europe. Wordsworth wrote of a fusing of "the round ocean and the living air, / And the blue sky and in the mind of Man." Beethoven would literally hug a linden tree in his backyard. He dedicated symphonies to landscapes and wrote, "The woods, the trees and the rocks give man the resonance he needs." Both men were speaking of a melding of inner and outer systems. It sounds a bit woolly, but they were auguring the explorations of twenty-first-century neuroscience, of human brain cells that sense environmental cues. Our nervous systems are built to resonate with set points derived from the natural world. Science is now bearing out what the Romantics knew to be true.

GROWING UP IN the dense, vertical habitat of a prewar apartment building, I was drawn to the verdant, magnetic acres of New York's Central Park. Starting in middle school, I went there most days and every weekend, riding a rusty Panasonic bike or walking, skating or sunbathing while tethered to a Walkman. We are animals, and like other animals, we seek places that give us what we need. Given the opportunity, children will decamp to tree houses and build forts, wanting spaces that feel safe but with easy access to open run-around areas. We work hard to make our homes and yards a certain way, and when we can afford to, we pay considerably more for residences or hotel rooms right on the beach, or the pastoral ninth hole, or a quiet, tree-lined street. We all want our starter castles on the corner of Prospect and Refuge. Experts tell us these habitat preferences are remarkably consistent across cultures and eras.

Yet until recently psychologists and neuroscientists didn't take these affinities very seriously. "Studying the impacts of the natural world on the brain is actually a scandalously new idea," Richard Louv, author of the 2008 bestseller *Last Child in the Woods*, told me. "It should have been studied thirty to fifty years ago." So why now? Probably because we're losing our connection to nature more dramatically than ever before. Thanks to a confluence of demographics and technology, we've pivoted further away from nature than any generation before us. At the same time, we're increasingly burdened by chronic ailments made worse by time spent indoors, from myopia and vitamin D deficiency to obesity, depression, loneliness and anxiety, among others.

In parts of East Asia, which suffers perhaps the greatest epidemic of indoor-itis, rates of nearsightedness in teenagers surpass 90 percent. Scientists used to attribute myopia to book-reading, but it instead appears to be closely linked to time spent living like naked mole rats, away from daylight. The sun primes the retina's dopamine receptors, and those in turn control the shape of the developing eye. We are learning about what this rift from the outdoors is doing to our retinal cells, but what about our minds?

We have gained much since the dawn of the Internet, but many experts argue we've also grown more irritable, less sociable, more narcissistic, more distracted and less cognitively nimble. We can't blame all our malaises on a separation from nature, but our complaints reveal some fraying of psychological resilience. There are times when we could all be a little less reactive, a little more empathetic, more focused and more grounded. That's where a nature dose can help, and many of the researchers in this book say they can prove it.

It wasn't apps or John Muir that compelled me to wonder about the relationship between nature and the human brain. For me, the exploration started when my husband accepted a job offer that would take us from an idyllic small mountain city to the hyper-urb of Washington, D.C. The summer evening we moved out of our house in Boulder, Colorado, was warm and clear. We stood on the curb watching the last of our dismayingly large pile of boxes, furniture and gear get tossed into an Atlas Van Lines truck. The kayaks were the last to go. Bright as jelly beans, scuffed by years of river rocks, they had no clue they were destined for a concrete parking pad in a big city.

Our next-door neighbors came out. Their kids draped arms around our kids. Soon small children from our branch of dead-end streets wandered up with their scooters and dogs. At ten and eight, our kids had been the elders, leading the pack to plastic-cup boat races in the irrigation ditch, raccoon spotting, tree climbing, rock painting and general mayhem among the shrubbery. There were days when they'd be outside from after school until dinner time, and I didn't really know exactly what they were doing.

The sky was pinkish. Never does Colorado look as beautiful as when lit by a summer sunset. I'm sure I was crying before the doors on the truck slid shut. Then my neighbor started and we were a couple of fools sniffling against the ornamental sage.

There were a lot of reasons I was sorry to be leaving the West, where we'd lived for two decades. Chief among them were my friends and colleagues, the kids' school and pals, our woodsy house, the mountains themselves. The trails near our house were ribbons of delight, filled with surprises like the baby scorpion who skittered across to say good-bye, the changing parade of wildflowers and my voluble hiker buddies as we dodged the clench-faced triathletes.

Even so, like a lot of people, I never really knew what I had until I lost it. What I didn't fully realize that evening the semi slid away with our worldly goods was how much the mountains had become my tonic. Nearly every day I was in them or on them or looking at them, often alone. Unlike a lot of people in Boulder, I was neither a seeker nor a fitness freak, so I didn't approach my walks with a quest for spiritual or material utility. And as a born and bred New Yorker, I don't use words like "tonic" lightly. I've never worn a heart monitor and clocked sprints or downloaded playlists from Olympic coaches. I would just head outside, usually walking, and if I couldn't get out, I'd get surly. When my feet were moving, I would think about whatever I needed to think about and the farther I went, the more I would space out. Sometimes I could by accident compose some writerly sentences in my head, or some insight might waft up, unbidden.

I'm not a wannabe mountain sprite. There's a lot I love about cities, like great cheap tacos and smart people in fantastic eyewear. It's just that I noticed some dramatic things about my mood, creativity, imagination and productivity in different environments, and I started to ponder it.

THE MOVING TRUCK pointed itself toward that anti-Arcadia that is the nation's capital, and we reluctantly followed. It was 104 degrees when we arrived, and my hair shriveled up into a pile of Brillo. This surely wasn't the East Coast; this was Manaus with suits. I ventured out to explore a nearby park early in the morning, and found that to get there, I needed to sprint across a highway and bushwhack along some bridge pilons to find the words "Pussy Fudge" waiting for me in spray paint. Our house was near a river but also near a major airport. Jets passed low overhead every sixty seconds. There was the noise, the smog, the gray, the heat. (To be fair, nature as well as civilization could wreck you here: the nonnative tiger mosquitoes as big as my thumbnail, the nymph deer ticks smaller than freckles. Both are capable of giving you diseases that can damage you neurologically and for life. Washington had names for weather events I'd never heard of or had to think about: derechos, polar vortices, level 4 hurricanes, heat-index advisories.)

I yearned for the mountains. And yearning is a devastating thing, because it is defined by loss. As the months ticked by, I realized that if I was going to explore what nature offers our brains, I also had to acknowledge what its absence means. I felt disoriented, overwhelmed, depressed. My mind had trouble focusing. I couldn't finish thoughts. I couldn't make decisions and I wasn't keen to get out of bed. I was perhaps, at least in part, suffering from what journalist Louv calls nature deficit disorder. (The DSM hasn't added it, but presumably they'd want to treat it with a pill.) Louv defines it as what happens when people, particularly children, spend little or no time outside in natural environments, resulting in physical and mental problems including anxiety and distraction. He also coined the tooth-some term "nature neurons" to highlight the essential link between our nervous systems and the natural world they evolved in. Was the breakage of this link really happening? Is there science supporting the notion of nature deficit disorder? If so, how much nature do we need to fix ourselves? Do we need to move into a hemlock tree like in a Jean Craighead George novel, or will looking out the window do?

If I was going to do more than merely survive in my new urban habitat, the type now shared by most people on earth, I was going to have to figure some things out. What was it about nature that people seem to need? How could we get enough of it in our lives in order to be our best selves? In the course of trying to answer these questions, I came to consider the human-nature connection on a neural level. Some weeks after we rolled into town, I left on assignment for Japan to write about an obscure and somewhat embarrassing Japanese practice called forest-bathing. There, I started to learn the science behind what I was experiencing at home. The Japanese researchers weren't content to leave nature to the realm of haikus—they wanted to measure its effects, document it, chart it and deliver the evidence to policy makers and the medical community. What the Japanese didn't really know, though, was why nature seemed to be helpful in alleviating so many things that ail us. And there were a lot of other things they didn't know: who was best helped, by what mechanisms in the brain and body, what was the right dose, and, moreover, what qualified as "nature"? I personally like Oscar Wilde's broad defini-tion: "a place where birds fly around uncooked."

Many scientists the world over are trying to find answers. My exploration of these questions would send me down a river in Idaho with a boatload of women veterans, to South Korea, where grown firemen hold hands in the woods, to sound labs measuring stress recovery, to treadmills in 3D virtual-reality rooms and to downtown Edinburgh, Scotland, where I'd walk around with a brain-measuring EEG unit wrapped around my scalp like a postmodern crown of thorns. I'd measure black carbon and my own blood pressure, pulse rate, cortisol and facial responses to "awe." I would meet researchers convinced that the secret to nature's power lies in its geometric fractal patterns, or its particular sound vibrations, or the aerosols from trees. It was a sensory extravaganza.

Scientists are quantifying nature's effects not only on mood and well-being, but also on our ability to think—to remember things, to plan, to create, to daydream and to focus—as well as on our social skills. There were times when I was skeptical, and times when I believed. I spent time with people who were trying to get well, people who were trying to get smart, people finding the best ways to educate young children (who are, by nature, exploratory, kinetic and full of wonder, all qualities enhanced by time outside) and people who were merely trying, like me, to stay sane in a frenetic world. Because of the two years I spent researching this book, I would emerge feeling better myself, and much more aware of the surprising science behind why I was feeling that way. And while "well-being" may sound like vague psychospeak, its impact is real. Enhancing it has been shown to add years to your life span.

I've divided the book into five parts to help make sense of the material, and to make it useful. The first part sets up the two dominant theories that attempt to explain why our brains need nature and that drive much of the research: the first chapter takes us to Japan, where researchers are quantifying nature's role in lowering stress and boosting mental health using a framework based on the biophilia hypothesis, the idea that we feel most "at home" in nature because we evolved there. The second chapter swerves over to Utah, where American neuroscientists are more interested in how nature helps restore our attention-addled brains to a state of sharper cognition. I've organized the rest of the book by nature dose. I explore the immediate effects of quick bursts, or "nearby nature" on our three main senses—smell, sound, sight. Then I look at what happens to our brains and bodies when we hang outside a bit longer to approximate the Finnish recommended nature dose: five hours a month. In Part Four, I take a deeper, longer dive into the wilderness, where really interesting things happen to our brains. This is where, in the words of neuroscientist David Strayer at the University of Utah, "something profound is going on." Finally, we'll look at what it all means to the way most of us live, in cities.

Throughout, there will be insights into how we can better construct our days, lives and communities so that everyone gains. Don't worry; I'm not going to tell you to pitch your smartphone over a waterfall. The world we live in is fully plugged in. But it's important to call out just how radically our lives have shifted indoors—and what those changes mean for our nervous systems—so that we may hope to ease and manage the transition.

My move to the city is a micronarrative of the demographic and geographical shifts occurring on a global scale. Homo sapiens officially became an urban species sometime in 2008. That's when the World Health Organization reported that for the first time more people throughout the world live in urban areas than rural ones. Last year in the United States, cities grew at a faster clip than suburban regions for the first time in a hundred years. Looked at another way, we are in the middle of the largest mass migration in modern times. Yet as humans shift their activities to cities, astoundingly little planning, resources and infrastructure go into making those spaces meet our psychological needs.

In Istanbul in the spring of 2013, eight people died and thousands were injured in protests stemming from the proposed paving-over of one of the last parks in the city, Taksim Gezi. Over 2 million of the region's trees had already been cut down to make way for a new airport and a new bridge over the Bosphorus Strait. The park was slated for a new shopping mall and luxury apartments. As bulldozers entered the park to mow down the urban forest, citizens blocked their way. They were willing to die for the last tree. "We will not leave until they declare the park is ours," said one twenty-four-year-old. (As of this writing, the trees still stand, but their fate remains uncertain.)

Taksim Gezi became a symbol not only of the importance of nature to city life, but to democracy itself, just as Frederick Law Olmsted knew all along. "A sense of enlarged freedom is to all, at all times, the most certain and the most valuable gratification afforded by a park," he wrote.

Yet we think of nature as a luxury, not a necessity. We don't recognize how much it elevates us, both personally and politically. That, ultimately, is the aspiration of this book: to find the best science behind our nature-primed neurons and to share it. Without this knowledge, we may not ever fully honor our deep, cranial connection to natural landscapes.

Not far from where I sent my lichen-rock photo into the Mappiness ether, two mighty rivers merge: the Green and the Colorado. It makes me happy to think of this geography because of a story of two goofy brothers I know, who, in college, built a raft out of inner-tubes and pallets, twisted out of their clothes, and pushed off the bank of the Green, heading to the confluence. They had baggies of gorp, a couple of jars of peanut butter, some water jugs. The water was calm on this stretch, and they were living the life. Just a couple of hours into the three-week trip, they got pulled over by a ranger. Fortunately, this was before the days of a required permit, fire pan and chemical toilet. But the naked boys were short one lifejacket. They were so busted. The ranger hauled them off to a county judge, who fined them, made them buy a lifejacket, and sent them back down the river (always better than being sent up the river). Those two guys are my brothers-in-law. This story has entered our sizable family canon of misadventures-by-uncles. But it seems ages ago that such a story would even be possible. Two college boys alone in the wilderness, having the time of their lives, able to make it weeks without civilization, minus a trip to a judge. Yet these two barely have gray hair; it was only a generation ago.

The dramatic loss of nature-based exploration in our children's lives and in our own has happened so fast we've hardly noticed it, much less remarked on it. "We evolved in nature. It's strange we'd be so disconnected," said Nisbet. Most of us don't know we're missing anything. We may have a pet and occasionally go to the beach, so what's the big deal? Well, what *is* the big deal? That's what I wanted to find out. And if something serious is missing, how do we recapture it?

As a journalist who writes frequently about the environment, I often end up writing about the way environment hurts our health, from flame retardants getting into human tissue to air pollution's effects on the developing brain. It was both a pleasure and a revelation to consider how, instead, our surroundings can also help prevent physical and mental problems and align us with the World Health Organization's definition of health: "a complete state of physical, mental and social well-being, and not merely the absence of disease or infirmity." The former health minister of Scotland calls this health-making "salutogenesis," inspired by the mid-twentieth-century sociologist Aaron Antonovsky, who asked, if the world is so crazy, what makes us able to keep sallying forth?

My city hair flattened to my scalp with gelatinous product, gulping vitamin D, I decided the answers are worth pursuing.

PART ONE

LOOKING FOR
NATURE NEURONS

The Biophilia Effect

In short, the brain evolved in a biocentric world.
—EDWARD O. WILSON

There is nothing you can see that is not a flower; there is nothing
you can think that is not the moon.
—BASHO

When I pictured *shinrin yoku,* "forest bathing," I con-
jured Sleeping Beauty in her corpse phase, surrounded
by primordial trees, twittering birds and shafts of
sunlight. You just knew she was somehow taking it all in, and she'd
awake refreshed, enlightened and ready for her hot prince. But this
was wrong on so many levels. First off, Japan doesn't have a lot of
primeval forest left, and second, you have to work at this, although
corpselike moments are not discouraged. In Chichibu-Tama-Kai
National Park, a ninety-minute train ride from Tokyo, I was supposed to be concentrating on the cicadas and the sound of a flowing creek when a loud Mitsubishi van rumbled by. It was disgorging
more campers to a nearby tent village where kids were running
around with their fishing poles and pink bed pillows. This was
nature, Japan-style.

The dozen others with me on our *shinrin yoku* hike didn't seem
to mind the distractions. The Japanese go crazy for this practice,

which is standard preventive medicine here. It involves cultivating your senses to open them to the woods. It's not about wilderness; it's about the nature/civilization hybrid the Japanese have cultivated for thousands of years. You can stroll a little, write a haiku, crack open a spicebush twig and inhale its woodsy, sassy scent. The whole notion is predicated on an ancient bond that can be unearthed with a few sensory tricks.

"People come out from the city and literally shower in the greenery," our guide, Kunio, explained to me. "This way, they are able to become relaxed." To help us along, Kunio—a volunteer ranger—had us standing still on a hillside, facing the creek, with our arms at our sides. I glanced around. We looked like earthlings transfixed by the light of the mother ship. Weathered and jolly, Kunio told us to breathe in for a count of seven seconds, hold for five, release. "Concentrate on your belly," he said.

We needed this. Most of us were urban desk jockeys. We looked like weak, shelled soybeans, tired and pale. Standing next to me was Ito Tatsuya, a forty-one-year-old Tokyo businessman. Like many day-hikers in this country, he carried an inordinate amount of gear, much of it dangling from his belt: a cell phone, a camera, a water bottle and a set of keys. The Japanese would make great boy scouts, which is probably why they make such great office workers, working longer hours than anyone else in the developed world. It's gotten to the point where they've coined a term, *karoshi*—death from overwork. The phenomenon was identified during the 1980s bubble economy when workers in their prime started dropping dead, and the concept reverberated into the future and throughout the developed world: civilization can kill us. Ito and I breathed in the pines and then dove into our bento boxes full of octopus and pickled root vegetables. Kunio was moving around, showing people the astonishingly twiggy walking-stick insect. Ito's shoulders seemed to be unclenching by the minute.

"When I'm out here, I don't think about things," he said, deftly scooping up shards of radish while I splattered mine onto the leaf litter.

"What's the Japanese word for 'stress'?" I asked.

"'Stress,'" he said.

WITH THE LARGEST concentration of giant trees in Japan, this park is an ideal place to put into practice the newest principles of Japanese wellness science. In a grove of rod-straight sugi pine, Kunio pulled a thermos from his massive daypack and served us some mountain-grown, wasabi-root-and bark-flavored tea. The idea with *shinrin yoku*, a term coined by the government in 1982 but based on ancient Shinto and Buddhist practices, is to let nature into your body through all five senses, so this was the taste part. I stretched out across the top of a cool, mossy boulder. A duck quacked. This may not have been the remote and craggy wilderness preferred by John Muir, but it didn't need to be. I was feeling pretty mellow, and scientific tests would soon validate this: at the end of the hike, my blood pressure had dropped a couple of points since the start of the hike. Ito's had dropped even more.

We knew this because we were on one of Japan's forty-eight official "Forest Therapy" trails designated for *shinrin yoku* by Japan's Forestry Agency. In an effort to benefit the Japanese and find nonextractive ways to use forests, which cover 68 percent of the country's landmass, the agency has funded about $4 million in forest-bathing research since 2003. It intends to designate one hundred Forest Therapy sites within ten years. Visitors here are routinely hauled off to a cabin to stick their arms in blood pressure machines, part of an effort to provide ever more data for the project. In addition to its government-funded studies and dozens of special trails, a small number of physicians in Japan have been certified in forest medicine. It's hard to overstate how unusual this is.

"The Japanese work is essential in my mind, a Rosetta stone," Alan Logan, a Harvard guest lecturer, naturopath and member of the scientific committee of the International Society of Nature and Forest Medicine (which is, naturally, based in Japan), had told me. "We have to validate the ideas scientifically through stress physiology or we're still at Walden Pond."

The Japanese have good reason to study how to unwind: In addition to those long workdays, pressure and competition for schools and jobs help drive the third-highest suicide rate in the world (after South Korea and Hungary). One-fifth of Japan's residents live in greater Tokyo, and 8.7 million people have to ride the metro every day. Rush hour is so crowded that white-gloved workers help shove people onto the trains, leading to another unique term, *tsukin jigoku*— commuting hell.

THE CIRCUMSCRIBED, urban life is of course not unique to Japan. I now reflected the nature-deprived trends myself. I spend too much time sitting inside. I maintain multiple social-media platforms that attenuate my ability to focus, think and self-reflect. Since moving to D.C., I've had crying jags in traffic jams, and at times I've been so tired I've had to pull over and nap on MacArthur Boulevard. When I do get out "in the woods," I seem to be doing it all wrong, forgetting or unable to hear the birds or notice any dappled anything. Instead, I grumble and obsess over my fate, my relationships and my kids' new schedules, which require military precision and Euclidean traffic calculations.

A couple of months after I moved, I told my new doctor I was feeling depressed. She did what general practitioners everywhere are doing and sent me off with a script for Zoloft. One in four middle-aged American women takes or has taken an antidepressant. One in fourteen children takes a drug for emotional or behavioral problems, reflecting about a fivefold increase since 1994. For me, as for a sizable percentage of others with mild depression, the meds didn't seem to work, and I hated the common side effects, which include everything from headaches to insomnia to low libido.

Moving on, I tried to grasp the destress crowd's favorite darling, meditation. The science is very convincing that it changes your brain in ways that make you smarter and kinder and generally less ruffled by life. The problem is, as with antidepressants, meditation doesn't work for many of us. Only 30 percent of aspirants are "fully adherent" after a standard eight-week course, according to Joshua Smyth, a biobehavioral psychologist at Pennsylvania State University. It has a high threshold to enlightenment.

But pretty much any slouching screen fiend can spend time in a pocket of trees somewhere. If there was one man who can demonstrate how forest therapy works, it's Yoshifumi Miyazaki. A physical anthropologist and vice director of the Center for Environment, Health and Field Sciences at Chiba University on the outskirts of Tokyo, he believes that because humans evolved in nature, it's where we feel most comfortable, even if we don't always know it.

In this, he is a proponent of a theory popularized by the widely revered Harvard entomologist E. O. Wilson: the biophilia hypothesis. It's been more or less appropriated by environmental psychologists into what's sometimes called the Stress-Reduction Theory or Psycho-Evolutionary Restoration Theory. Wilson didn't actually coin the word "biophilia"; that honor goes to social psychologist Erich Fromm, who described it in 1973 as "the passionate love of life and of all that is alive; it is the wish to further growth, whether in a person, a plant, an idea or a social group."

Wilson distills the idea more precisely as residing in the natural world, identifying "the innately emotional affiliation of human beings to other living organisms," as an evolutionary adaptation aiding not only survival but broader human fulfillment. Although no specific genes have been found for biophilia, it's well

recognized—ironically, some from studies of bio*phobia* or fear—
that even today our brains respond powerfully and innately to natural stimuli. One powerful example: *snake!* Our visual cortex picks
up snake patterns and movements more quickly than other kinds
of patterns. It's likely that snakes even drove the evolution of our
highly sensitive depth perception, according to University of California anthropologist Lynne Isbell. She discovered special neurons
in the brain's pulvinar region, a visual system unique to humans,
apes and monkeys. Primates who evolved in places seething with
venomous snakes have better vision than primates who didn't
evolve in those places.

But survival wasn't only about avoiding harm. It was also about
finding the best food, shelter and other resources. It makes sense
that certain habitats would trigger a neural bath of happy hormones, and that our brains would acquire the easy ability to "learn"
this in the same way we learn to fear snakes and spiders. Going
beyond that, our ancestors also had to learn how to recover from
stress, Pleistocene-style. After they were chased by a lion or dropped
a precious tuber over a cliff, they had to get over it in order to be wel-
comed back to the tribe, without which there was little survival. The
biophilia hypothesis posits that peaceful or nurturing elements of
nature helped us regain equanimity, cognitive clarity, empathy
and hope. When love, laughter and music weren't around, there was
always a sunset. The humans who were most attuned to the cues of
nature were the ones who survived to pass on those traits. Biophilia
explains why even today we build houses on the lake, why every child
wants a teddy bear, and why Apple names itself after a fruit and its
software after noble predators, surfing spots and national parks.
The company is brilliant at instilling biophilic longing and affiliation at the very same time it lures us inside.

It should come as no surprise that crosstalk operates between
the brain and nature, but we're less aware of the ever-widening gulf

between the world our nervous systems evolved in and the world they live in now. We celebrate our brains' plasticity, but plasticity goes only so far. As Miyazaki explained it, "throughout our evolution, we've spent 99.9 percent of our time in nature. Our physiology is still adapted to it. During everyday life, a feeling of comfort can be achieved if our rhythms are synchronized with those of the environment." Of course, he's talking about the nice parts of nature found in the hillsides of Japan, not the pestilential scum ponds or barren terrains of the globe that also constitute nature. Stick an office worker there, and relaxation will likely not be happening. But Miyazaki points out that naturalistic outdoor environments in general remain some of the only places where we engage all five senses, and thus, by definition, are fully, physically alive. It is where our savanna-bred brains are, to borrow from John Muir, "home," whether we consciously know it or not. By contrast, Muir wrote of time not in the wilderness: "I am degenerating into a machine for making money." Make that a machine with clogging pipes.

To prove that our physiology responds to different habitats, Miyazaki's taken hundreds of research subjects into the woods since 2004. He and his colleague Juyoung Lee, then also of Chiba University, found that leisurely forest walks, compared to urban walks, deliver a 12 percent decrease in cortisol levels. But that wasn't all; they recorded a 7 percent decrease in sympathetic nerve activity, a 1.4 percent decrease in blood pressure, and a 6 percent decrease in heart rate. On psychology questionnaires, they also report better moods and lowered anxiety.

As Miyazaki concluded in a 2011 paper, "this shows that stressful states can be relieved by *shinrin* therapy." And the Japanese eat it up, with nearly a quarter of the population partaking in some *shinrin* action. Hundreds of thousands of visitors walk the Forest Therapy trails each year.

I MET UP WITH Miyazaki at the country's newest proposed therapy site, Juniko state park on the edge of the Shirakami Mountains in northern Japan. He was swatting mosquitoes from his face and neatly trimmed gray hair. He wasn't looking relaxed at all. It had rained recently, and he was worried the trail might be too muddy for his upcoming walking experiment. He was kicking some rocks out of the way and overseeing the setting-up of a netted, canopied minilab. The next morning, Miyazaki and Lee would be bringing twelve male college-student volunteers here, measuring various vital signs after they walked and sat and generally forest-bathed. Then they would repeat the experiment the next day in downtown Hirosaki, a city of 100,000, two hours away by car. I would join as one of Miyazaki's guinea pigs.

The trail deemed walkable, several of us retired to a quiet restaurant in Hirosaki. We took off our shoes and sat cross-legged on the floor while Miyazaki ordered and then distributed a baffling array of dishes involving goopy eggs, gelatinous balls and surf-and-turf combinations.

"Why do the Japanese think about nature so much?" I asked Miyazaki, who was preparing to eat a manta ray.

"Don't Americans think about nature?" he asked me.

I considered. "Some do and some don't." But I was thinking, an amazing amount of us don't, given our downward trends in outdoor time and visits to parks.

"Well," he mused. "In our culture, nature is part of our minds and bodies and philosophy. In our tradition, all things are relative to something else. In Western thought, all things are absolute."

Maybe it was the sake, but he was losing me.

"The difference is in language," he continued. "If I ask you, 'Is a human a dog?' you say, 'No, a human is not a dog.' In Japan, we say, 'Yes, a human is not a dog.' The great sensei of nature research peered at me over his chopsticks. I was reminded of the story of the

Zen student who asks his teacher, "How do you see so much?" and the teacher responds, "I close my eyes."

Miyazaki's answer, I understood, was like a koan, tantalizing and confounding at the same time. But you had to trust the guy was onto something.

THE NEXT MORNING, the college boys and I took turns sitting in the mobile lab at the trailhead. We placed hard cotton cylinders under our tongues for two minutes, then spit them out into test tubes. That would record our levels of cortisol, a hormone made in the adrenal cortex. We got hooked up to probes and devices. The team was inaugurating a brain-measuring, battery-powered, near-infrared spectrometer that, when deployed, gave me a sensation of leeches sticking to my forehead. We'd repeat all these measurements at the end of the walk and again in the cityscape.

To gauge our physiological responses to these environments, Miyazaki and Lee look at changes in blood pressure, pulse rate, variable heart rate, salivary cortisol and, new this year, hemoglobin in the brain's prefrontal cortex. When aggregated, these metrics paint a picture of our bifurcated nervous system. When we are relaxed and at ease in our environment, our parasympathetic system—sometimes called the "rest and digest" branch—kicks in. This is why food tastes better in the outdoors, explains Miyazaki. But the demands and constant stimuli of modern life tend to trigger our sympathetic nervous system, which governs fight-or-flight behaviors. And trigger it, and trigger it. We suffer the consequences: a long trail of research dating back to the 1930s shows people who produce chronically high cortisol levels and high blood pressure are more prone to heart disease, metabolic disease, dementia and depression. More recent research shows that the steady stress of urban living changes the brain in ways that can increase our odds of schizophrenia, anxiety and mood disorders.

When it was my turn to wander through the forest for fifteen minutes, I was happy to break free from the wires. The loud pulse of cicadas echoed through the woods. Light filtered gently through the beeches and Japanese horse chestnuts and the earth smelled like good damp dirt. An elderly couple ambled by, assisted by walking sticks and a bear bell. I was briefly mesmerized by a yellow butterfly. I could see why Juniko, a leafy network of trails and lakes, is a candidate for the country's next forest therapy station. Local and park officials are seeking the designation because where there's forest therapy, there are tourists and their yen. Miyazaki may have a mystical side, but what drives him is more data. It's a convenient arrangement.

The Japanese work on physiology and the brain takes advantage of new tools of brain science, but it builds on decades of psych-talk about the health benefits of being in nature. Miyazaki wasn't the first to record physical stress recovery in nature. A young psychologist named Roger Ulrich was curious why so many Michigan drivers chose to go out of their way to take a tree-lined roadway to the mall. In 1986, using the expensive and cumbersome equipment of the time, he hooked up an electroencephalograph (EEG) unit to the heads of healthy volunteers while they viewed slides of nature scenes or utilitarian urban buildings. The subjects assigned to nature showed higher alpha wave activity, a wavelength associated with relaxation, meditation and increased serotonin. In another experiment, he stressed out 120 students by showing them movies of bloody accidents in a woodworking shop. He knew they were distressed because he measured their sympathetic nervous activity—the sweat glands on their skin, their heart rates and their blood pressure. Afterward, some students were assigned to watch a ten-minute video of nature scenes and some to watch videos of urban scenes, from a pedestrian mall to cars on a road. The results were dramatic: within five minutes, the brains-on-nature returned to baseline. The brains-on-built-environment recovered only partway—as indicated by those nervous system measures—even more than ten minutes later.

Despite early promise, the study of brains-on-nature went fairly dark for a couple of decades. It was considered soft science, much of it based on qualitative measures in a medical world dazzled by genetics and modern chemistry and funded by pharmaceutical companies that didn't stand to make a profit from houseplants or garden views. The renewed interest of late represents a convergence of ideas and events: the relentless march of obesity, depression and anxiety (even in affluent communities and despite more medication), the growing recognition of the role of the environment on genes, and the growing academic and cultural unease with our widening breach from the outdoors.

NOT SURPRISINGLY, MY urban peregrination wasn't quite as pleasant as the soft green trail of Juniko. Downtown Hirosaki is far less green than D.C. There are transit stations, shops selling basic goods, and people on the go. In the height of summer, the asphalt was baking. Shoppers rushed in and out of a department store whose busy windows advertised "spaghetti with tomato cream." I passed four parking lots, two taxi stands, a bus station, and two loudly idling buses belching fumes. My nervous system responded. My systolic blood pressure had dropped six points after walking in the forest. It went up six points after walking in the city. Which of course begs the question: How long do the feel-good effects of nature last? Do they just get wiped out by the first traffic jam or cell phone tone?

Miyazaki's sometime collaborator, an immunologist in the department of environmental medicine at Nippon Medical School in Tokyo, wondered the same thing. Qing Li is interested in nature's effect on mood states and stress as manifested in the human immune system. Specifically, he studies natural killer immune cells, called NK cells, which protect us from disease agents and can, like cortisol and hemoglobin, be reliably measured in a laboratory. A type of white blood cell, they're handy to have around, since they send self-destruct messages to tumors and virus-infected cells. It's been known for a long time that factors like stress, aging, and pesticides can reduce your NK count, at least temporarily. So, Li wondered, if nature reduces stress, could it also *increase* your NK cells and thereby help you fight infections and cancer?

To find out, Li brought a group of middle-aged Tokyo businessmen into the woods in 2008. For three days, they spent a couple of hours each morning hiking. By the end, blood tests showed their natural killer cells had increased 40 percent. Moreover, the boost lasted for seven days. A month later, their NK count was still 15 percent higher than when they started. In contrast, during urban walking trips of the same duration, NK levels didn't change. Since then, Li has published results from similar studies with male and female subjects in half a dozen peer-reviewed journals. In one, Li was curious to know if a one-hour trip to a city park would have a similar effect, since most of us can't spend three days a week walking in the woods. It did, although the immune surge didn't last quite as long.

What was going on? Li suspected the trees. Specifically, he wondered if NK cells are boosted by "aromatic volatile substances," otherwise known as nice tree smells, and sometimes called phytoncides. These are the turpenes, pinenes, limonenes and other essential oils emitted by evergreens and many other trees. Scientists have identified over a hundred of these phytoncides in the Japanese countryside, and virtually none in city air that's not directly above a park. This wasn't a totally left-field idea. Since at least 2002, studies have attributed healthful properties to soil compounds like actinomycetes—which the human nose can detect at concentrations of 10 parts per trillion—and of course we harvest mold spores to

make critical antibiotics like penicillin. Dirt can heal: in two separate experiments in England and the United States in 2007 and
2010, the mice lucky enough to be exposed to a common soil bacterium, *Mycobacterium vaccae*, performed better in a maze, showed
less anxiety and produced more serotonin, a neurotransmitter
many scientists think is associated with happiness.

To test the phytoncide theory, Li locked thirteen subjects in
hotel rooms for three nights. In some rooms, he rigged a humidifier to vaporize stem oil from hinoki cypress trees, which are common in Japan; other rooms emitted eau-de-nothing. The results?
The cypress sleepers experienced a 20 percent increase in NK cells
during their stay, and they also reported feeling less fatigued. The
control group saw no changes.

"It's like a miracle drug," said Li, when I interviewed him at his
university lab in Tokyo.

It sounds totally hokey, even unbelievable, that evergreen scents—
not unlike the thing that dangles from taxicab rear-view mirrors—
could help us live longer. But Li found similar results with NK cells
exposed to phytoncides in a petri dish. The cells increased, and so
did anticancer proteins and proteases called granulysin, granzymes
A and B and perforin, which cause tumor cells to self-destruct. It's
unclear whether there's something magical in the aromatic molecules or if the smell simply makes people feel good, reducing stress.
Li's olfaction theory is unconventional, but it contains some of that
zen five-sense wisdom. While American researchers are mostly
showing people pictures of nature or sending them out for loops
around the campus green, the ones in Japan are practically pouring
it into every orifice.

Li, the chairman of the Japanese Society of Forest Medicine,
uses some of his insights in his own life. "In fact," he said, "I use a
humidifier with cypress oil almost every night in the winter!" You

don't need to harvest your own; he said standard health-store aro-matherapy oils should do the job.

"What else do you recommend?" I asked the middle-aged man with the bowl haircut.

Clearly, Li gets asked this a lot. He had a small list. "If you have time for vacation, don't go to a city. Go to a natural area. Try to go one weekend a month. Visit a park at least once a week. Gardening is good. On urban walks, try to walk under trees, not across fields. Go to a quiet place. Near water is also good."

I could see my morning walk back in D.C. transforming before my eyes.

I COULDN'T HELP wondering, though, if having more data on how nature changes our brains and immune cells would actually lure more of us into the woods. We also know we're supposed to eat more leafy greens, but most of us don't. The kale analogy is pretty apt, because it turns out that even when we don't like nature, such as during lousy winter conditions, it ends up benefiting us. At least that's what University of Chicago professor Mark Berman found when research subjects took walks in an arboretum during a blus-tery winter day. The walkers didn't enjoy themselves, but they still performed better on tests measuring short-term memory and atten-tion. We'll learn more about his work in the next chapter.

While the Japanese researchers understand our draw to nature, many American ones seem preoccupied with our pull away from it, our distractions, inertia and addictions. They want to know if resisting that pull and turning toward nature can enhance our productivity. Perhaps this cultural difference is what Miyazaki was explaining over his plate of sting ray: oneness versus me-ness. Americans want to know what can nature—that stuff over there—do for us? More Beowulf than Basho, the Americans want to slay the dragon and get back to the mead hall. They prefer to use delineated spurts of nature to optimize their success. Maybe they can even use digital nature and forget the bugs and rain altogether.

I would head back to the States, to Utah, to see what some American researchers were up to and how they were preparing to tackle the research. Their inquiries, geared to cognition and creativity, provide the other main theoretical framework for understanding how nature acts on our brains. In the meantime, I would be scratching and sniffing some pinecones. The bark tea? Not so much. Running my hands through the moss, sure.

Why not? After all, yes, I am not a dog.

How Many Neuroscientists Does It Take to Find a Stinking Milkvetch?

We used to wait
We used to waste hours just walking around
—ARCADE FIRE

When you head for the desert, David Strayer is the man you want behind the wheel. He never texts or talks on the phone while driving. He doesn't even approve of eating in the car. A cognitive psychologist at the University of Utah's Applied Cognition Lab, Strayer knows our brains are prone to mistakes, especially when we're multitasking and dodging distractions. The country's foremost expert on this topic, he frequently briefs Congress on the dangers of cell-phone use, which his research has shown to be as detrimental to driving ability as alcohol. He has recently begun to take on voice-recognition technology, like Siri and the computers that come with virtually all new autos.

"I talk to Siri all the time!" I said from the backseat of Strayer's 4Runner, my phone and its riveting Mappiness app in my pocket.

"Don't talk to Siri!" He implored me and the others in the SUV. Apple is very miffed at Strayer. So are GM and Ford.

For all his expertise with automobiles, Strayer has taken on their opposite in his latest line of inquiry: nature. As a longtime river rafter, backpacker and hiker, he knows he gets his best ideas in the wilderness. Now he wants to know why.

Buddha, Jesus, and Reese Witherspoon all went to the desert to seek wisdom. David Strayer was following the pattern, and bringing a half-dozen neuroscientists with him. Their plan: to figure out how to study the effect of something as beautiful and complex as nature on something as beautiful and complex as the human brain. While the Japanese begin with the premise of biophilia—our innate emotional connection to living things—and Mappiness assesses feelings, this group was all about cognition. Strayer's team was less interested in amorphous concepts of well-being and more interested in watching and measuring how nature helps us think, solve problems and work together. Results should be controlled, imaged, measured, charted, recrunched, replicated, regressed into chi squares and attacked in multiple studies from unexpected angles. On this retreat, they would need to come up with questions and experimental designs that could pass muster with their peers, and with themselves.

But for now, it was time to engineer some decent hiking. Strayer had invited the group to Moab, a scruffy town of mountain bikers and river runners in southern Utah named after an ancient kingdom. With its proximity to outrageous scenery—as well as to purveyors of decent 3.2 ale—it seemed a fitting place to discuss and plan experiments for assessing nature's impact on the brain. Strayer was the George Clooney to this *Ocean's Eleven* of nerds trying to crack a scientific problem. He had the maps, the supplies and the funding from the National Academies of Science. For my part, I wanted to understand where the neuroscientists were coming from theoretically and to learn about their doubts and biases as I set off on my own exploration about nature and health. Sitting in the car with me that first day were Lisa Fournier and her husband, Brian Dyre, psychologists from two universities near Pullman, Washington. Dyre was the biggest doubter in the group.

"I'm a skeptic about the restorative effects of nature," he told me. "I believe people feel good but I wonder about the mechanism—is it that you're just away from daily cares and is the benefit that you're in a new mind-set? Is it just a cheap and easy way to get to a new mind-set?" Dyre thinks being in nature might be no different from playing music or visiting a museum. The experience is diverting, pleasant and sometimes social. Period.

And in fact, science has shown that those things—music, friends, cultural events—are good for our mental health. Why should there be something superior about nature? Maybe a bunch of tree huggers just want that to be the case. It would be more fodder for their progressive agendas—more parks and wetlands, fewer paved megadevelopments and corporate theme parks. Museums, bands, legions of friends: they tend to be found in cities.

Skeptical or not, Dyre liked the scenery. We started out in Arches National Park, walking toward a landform called the Double O Arch via a series of red, slickrock fins characterized by steep sides and expansive views. It was a bit like walking on a dragon's back. A wooden sign warned CAUTION: PRIMITIVE TRAIL DANGEROUS HIKING. I was loving it. To arrive here from D.C. and inhale the desert was like climbing out of a basement. Everywhere was sky and light and the unlikeliest colors and collections of wind-worn twisted rock. It was a visual feast.

After picnicking on a narrow tongue of rock, we found the remarkable double-decker arch, which looked something like a giant bracelet of rock set atop a lifesaver. A few of us gingerly climbed atop the delicate upper bracelet. From the top, the world fell away on two sides. It felt treacherous, in a good way. Below us we could see a reclining Adam Gazzaley, who is an avid photographer when he's not authoring lead neuroscience articles for the journal *Nature*. We posed for some snapshots and got out of there.

"I just had this amazing thing happen," said Gazzaley when we reached him. "I was lying there, trying to get a shot of my feet and the rock and the sky, and all of a sudden I figured out something I never figured out before. I could take a vertical panorama! From bottom to top!" Gazzaley was now giggling. He showed us the tiny vertical panorama on his phone, but between the glare and the size it was not much to see.

"Half a day in nature and you're already more creative!" I said.

"I know, right?"

THIS WAS DAVE Strayer's third neuroscientists-in-the-desert con-fab. The first took place in 2010, a thirty-two-mile backpacking trip in Grand Gulch. After that came a five-day river trip with a slightly larger group. A canoe tipped over, two esteemed neuroscientists fell out, and a photographer from the *New York Times* caught it on camera. It was all a little embarrassing. The point of that river trip was for Strayer to infect his colleagues with his somewhat eccentric ideas having to do with the creativity and peace that are unleashed when you take off your watch, turn off your devices, and head into the wild. Of this group, Strayer is the one who buys most into the Power of Nature. But he knew he needed the street cred and technical lab expertise of the others.

The plan worked well enough. After five days, the scientists were uncannily relaxed, some more so than they had been in years, and they agreed to test Strayer's ideas. They came up with a pilot study to measure the creativity of fifty-six Outward Bound participants. Half took a test called the Remote Associates Test before the trip; half did so after three days of hiking. A fun and challenging measure of intuition and "convergent creativity," the RAT gives you three words and asks you to come up with a word that links them

(like water/tobacco/stove: answer—pipe. Here's a harder one: way/
ground/weather: answer is in the footnote.* If you can't guess it,
go stare at a tree and try again. Hint: it is not "under"). Although
it was a small study, the results (published in PLoS ONE) blew the
researchers away: a 50 percent improvement in creativity after just
a few days in nature.

Fifty percent! Who wouldn't want to harness that power? But it
needed to be replicated and teased apart. So Strayer chased down a
new grant, enough money to bring everyone together here and eventually run a couple of larger, more ambitious studies with the input
of the group. On this trip, the scientists were staying in a hotel,
albeit with a fire pit on a roofdeck. It was a compromise between
convenience and cave-dwelling. The plan was to hike and run rivers
during the day, sit around the fire at night and brainstorm experimental design. Drinks included.

Even though the Outward Bound study was intriguing, there
were a lot of variables going on and plenty of reasons to be wary of
the findings. Was it "nature" that improved performance, or was
it hanging out socially in a stimulating group for several days?
Was there simply a brightening of mood that made people sharper,
perhaps caused by better sleep, or the surprisingly good pow-
dered lentils (okay, unlikely), or a flirtation with the rock-climbing
instructor? The notion of "nature experience" could be exceedingly
difficult to unpack. "I think there's a recalibration of your senses, of
seeing and noticing," said Strayer. "I'd like to have empirical data to
assert or refute that hypothesis."

THANKS TO THE grant money, the scientists were able to dine a few
steps up from freeze-dried hummus. The first night after Arches,
they headed to Moab's finest (and only) Thai restaurant. Art Kramer,

---

* The answer is "fair."

a neuroscientist, had arrived from the University of Illinois, where he directs the Beckman Institute for Advanced Science and Technology. In his early sixties, he's clearly the senior yoda of the group. He greeted us and dove into the pad se. Smallish and solid, he's a man who gives the impression of intensity in all pursuits. "He talks at squirrel speed," one of the others had warned me. At one time or another, nearly everyone here (except Gazzaley) studied with him or worked in his lab. Strayer was his first doctoral student, back when they were studying pilot error. Kramer has always been fascinated by how humans learn skills and what makes them screw up. He's consulted for the military, NASA and the Federal Aviation Administration, among others.

But what Kramer is really known for—indeed, famous for, in the world of neuroscience—is showing how exercise protects the brain from cognitive decline in aging. Among his dozens of influential studies are those showing that exercise causes new brain cells to grow, especially in areas related to memory, executive function and spatial perception. Before Kramer's work, no one really believed physical activity could lead to such clear and important effects. Now people everywhere are routinely told that exercise is the single best way to prevent aging-related cognitive decline. Kramer's studies helped change the way the profession and society think. They are what scientists dream of.

"In 1992, the exercise/brain literature was where the nature literature is now," said Strayer. "My goal in the next ten years is to do for nature what he did on exercise and cognition."

If you draw a Venn diagram of the scientific interests of everyone around the vinyl-draped dinner table, the circles would overlap over one central theme: attention. Other scientists studying the effects of nature may be interested in other things, like emotional regulation, or stress, or the immune system. But in Team Moab's worldview, attention is the lingua franca from which all mental states spring. I'd be hearing a lot more about it.

Kramer sipped a lassi and briefly checked his phone. I asked him if he would be following Strayer's advice and taking a three-day tech break while in Moab. He peered at me rather severely.

"I brought four computers." He paused. "I can do it though. I lived in a snow cave for a month." Several heads swiveled his direction. "He's a sensation-seeker," Strayer explained.

"Definitely," Kramer said.

"Do you still have your Harley?" someone asked.

"Yep." Kramer pulled up a photo of a red motorcycle on his phone.

"Still wearing leather?" asked Strayer.

"Yeah, a jacket."

"No pants?"

"Well, I always wear pants."

WE WERE READY to experience some of the benefits of tech withdrawal in a place with no cell service. For the next day's hike up Hunter Canyon, Gazzaley planned to ditch his phone altogether, pulling out a beloved Real Camera. I expressed an interest in identifying wildflowers. Without the Internet, I'd need to go old-school: a laminated flower guide presented to me that morning by Ruth Ann Atchley, a psychologist from the University of Kansas. It's worth noting that she and her husband, Paul Atchley, who is another expert in distracted driving, managed to hold off owning smartphones until several weeks earlier, and then only to help manage email while traveling. These two are definitely not playing Crossy Road.

As we waited for the others to gather in the lobby, Paul Atchley wondered aloud if the restorative benefits of nature might in fact spring from what's not outside: the pings and dings and mental disruptions of a wired life. It was part of the ongoing conversation about which factors to isolate in upcoming studies.

"Is the explosion of attractive technologies that give our brains social interactions negatively impacting us, and is the cure to go back to an environment that our brain resonates with?" He answered his own question. "Tech is leading us in a negative direction and nature may prevent that." Both Paul Atchley and Strayer have been heavily influenced by the research of the late Stanford sociologist Clifford Nass. His well-regarded studies show that heavy media multitaskers have an impaired ability to focus on cognitively demanding tasks. Furthermore, his study of 2,300 preteen girls showed those with the highest rates of media use were less developed socially and emotionally than their peers. (Sadly, for Nass, healthful nature was not the antidote; the fifty-five-year-old died just after taking a hike.)

"Remember that guy at the Met who was talking on his cell phone and actually leaned against a Jackson Pollock?" continued Paul Atchley, shaking his head.

"Does less nature and more technology change who we fundamentally are?" asked Strayer.

"Hey, I'm alive because of technology," interjected Kramer. "I take statins, and I'm alive."

"I really mean phones, TV, digital media," said Strayer. "They're stimulating and flashing and probably addictive."

Paul Atchley was warming up. "Thirty-six percent of people check their cell phones while having sex. Seventy percent of people sleep with their phone."

Strayer: "The average person looks at their phone 150 times day. The average teen sends 3,000 text messages a month. These are hallmarks of an addictive, compulsive personality. We're wired to have social connection, to sit around the campfire, face-to-face. Social connection is like sugar."

Ruth Ann Atchley felt the need to reel them in. Passing out sunscreen, she was part hostess, part mediator. "Yes but what is it about nature?" she asked her husband.

"You see," she explained, looking at me, "he argues about getting away from tech and I argue about being in this space. I'm all Disney

movies and he's *House of Cards*. He thinks people's nature is negative." Paul shrugged but didn't disagree. "My hypothesis," she continued, "is when you're engaged in nature, it leads to mindfulness.
It's passive, the world is coming and going. It's so good for depression. When you walk out in nature, it's like wearing rose-colored
glasses. In nature everything is a little more positive, there's a little
more connectedness. This is the world in which we are supposed to
be. Plus, most of us have positive memories of childhood in nature."

Gazzaley, having arrived, now jumped in. "Well, in nature I do
feel more relaxed more quickly than anywhere else, but I didn't
spend time in nature as child." He grew up in Rockaway, New York,
riding the subway four hours a day to and from the Bronx High
School of Science. "By lunch yesterday, I was definitely relaxed."

Lisa Fournier, who had also joined us, roused: "That's affirming
the consequent. We're biased, we're just affirming our beliefs, and
the experiments reflect that."

Ruth Ann Atchley: "You don't go onto Outward Bound unless you
already believe it's helping you. But they had no idea what we were
looking for (in the cognitive tests)."

Fournier: "The placebo effects are so strong."

Kramer: "We're all skeptics."

Paul Atchley, hoisting his daypack: "I'll cite the *X-Files*. I want to
believe."

AND SO THE skeptics and the believers marched out of the Best
Western. I drove to the trailhead with Paul Atchley and Strayer. As
the strange, folded landscape revealed itself, I found myself wondering about the significance of attention, and its role in why nature
makes us smarter, as Strayer contends. Psychologists have been fascinated by the concept of attention for a long time, although it's now
enjoying a resurgence in our age of distraction, or what Paul Atchley
has called "the attention economy."

Attention is our currency, and it's precious. William James, the philosopher, pioneering experimental psychologist and brother of Henry James, devoted an entire chapter of his classic *The Principles of Psychology* to attention, published in 1890. In it, he writes, "Every one knows what attention is. It is the taking possession by the mind. . . ." and "My experience is what I agree to attend to . . . Without selective interest, experience is an utter chaos." Notably, James divided attention into two basic types that continue to define the way we think about it: voluntary, active attention (such as when we attend to tasks) and involuntary or reflex attention, as when something demands our focus, like a noise or sound or play of light or even a wayward thought. Decades before text alerts, philosophers were concerned by what James refers to as the "confused, dazed, scatter-brained state which in French is called distraction." (Before I leave James, I can't resist mentioning that he suffered from depression and experienced a transformative experience while hiking in the Adirondacks in 1898. As he wrote to his wife, he "got into a state of spiritual alertness of the most vital description." Emerson was his godfather, so perhaps he was primed to attend voluntarily to this possibility.)

James knew that staying on task was hard, hard work, and that without this ability, as Nass confirmed, we become dumber, at least by certain measures (by other measures, the distractions of the digital age may be a reasonable trade for what our brains gain in access to more information and more memory storage). But interestingly, we're also limited in our ability to take in our surroundings, because otherwise our brains would be overwhelmed by stimuli. Our field of vision is surprisingly narrow; our hearing isn't great either, and most of what we hear and "see" we don't actually process at all. We get on in the world because our brains are pretty good at automatic triage.

"Most of the time your brain can filter things out," said Strayer, driving the black 4Runner over an increasingly rough dirt road. "It's a strategic process. If traffic is heavy, your brain literally stops listening to NPR. Radio is a passive signal, but talking is a whole different thing, and if you're on the phone talking to your spouse, that's more difficult to shut out." Hence your inability to respond as quickly as you should to traffic signals, signs and pedestrians. Social information, as all Tweeters, texters and emailers know, draws our attention and is tough to shut out. I was reminded of a funny automated email response sent by a scientist on vacation (which I learned about through, of course, Twitter): "I am away from the office and checking email intermittently. If your email is not urgent, I'll probably still reply. I have a problem."

"Attention is everything," explained Paul Atchley, pivoting around in the front seat. "Without it, we don't see, hear, taste. Your brain keeps track of about four things at once. How do you prioritize what's important and what's not? Through inhibition. I've always found it interesting that most connections in the brain are inhibitory functions. We have far more information than we can deal with. Most of what the brain is doing is filtering, tuning stuff out so we can focus in on things that are relevant."

Because of this interplay of observation, selective attention and inhibition, humans are able to achieve higher-order cognition, which includes creative problem-solving, goal-following, planning and multitasking. The problem is that all this inhibition and filtering uses up cognitive fuel. It wallops us. As Stanford neuroscientist Daniel Levitin points out in *The Organized Mind*, our brain's processing speed is surprisingly slow, about 120 bits per second. For perspective, it takes 60 bits per second just to understand one person speaking to us. Directed attention, or voluntary attention, is a limited resource. When it flags, we make mistakes; we get irritable. Moreover, task-switching, which is something we do an awful lot of these days, burns up precious oxygenated glucose from the prefrontal cortex and other areas of the brain, and this is energy we need for both cognitive and physical performance. It's no wonder it feels pretty good to space out and watch a butterfly. Of course, that requires brain real estate too, but it's different real estate, and that's a key point.

As we neared the trailhead, the brilliant sky contrasted dramatically with the red cliffs through the front window. A corridor of green creek bed emerged from a seam in the landscape. "From my perspective," Atchley continued, sweeping his hand across the view, "what this environment is doing to us right now is giving us fewer choices. And by having fewer choices, your attentional system functions better for higher-order things. In the office environment, you've got emails, alerts, sounds. That's a lot of filtering and so it's harder to think deeply. Here the filtering requirements are not demanding so you have the capacity to focus on deeper thought."

COMING INTO THIS project, I believed that being in spectacular or even just pleasant natural environments helps me destress, think more clearly and feel grounded in a way that made me a better person. But I found myself resisting the idea that our Pleistocene ancestors had it so much better. Here in Moab were a bunch of middle-aged scientists who disliked their cell phones and saw the effect phones were having on their undergraduates, many of whom were distractible, listless and anxious. But it seemed too convenient and ahistorical to think that our modern stressed-out lives are somehow worse than the stressed-out lives of our forebears. I worried that the nature justifiers might be overly romanticizing cavemen (especially the men) who presumably got to skip across the veldt stalking game, building up their deltoids and engaging in bro rituals by the light of a crackling fire. But, hello. Hunter-gatherer child mortality rates alone would have sent most families into extreme grief, not to mention the dire uncertainties of food, weather and territorial warfare.

Humans have brains that are sensitive to social and emotional stress and we always have. Perhaps what matters is not the source of the stress but the ability to recover from it. This is a key point, because it's perhaps what we've lost by giving up our connection to the night skies, the bracing air and the companionate chorus of birds. When I'm walking across a pleasant landscape, I feel I have time and I feel I have space. I'm breathing deeply things that smell good and seeing things that bring delight. It's hard not to feel the pull of a grounded reality when you're dipping into a muddy trail or a flowing river. Speaking of which, we finally parked the vehicles and formed into loose walking pairs as we joined up with the creek path. The trail was sandy, the sky blue, and a gentle breeze rustled the sedges and stalks at our feet.

Up ahead, I came upon Kramer. His life of adventure had caught up with him. He wore a brace on his left knee (a high-speed skiing accident) and walked with a limp, but he walked fast. He will never be the type to watch the moss grow. He told me stories of nearly succumbing to dehydration in the Tetons and braving treacherous river crossings in Alaska. When he was ten years old and growing up in New York, he was conscripted into an elite division of scouts called the Order of the Arrow. He was given a knife, one egg and a fire-starting kit and sent off to the woods, alone, for three days. He has no doubt these experiences have helped him in life, but for him, it wasn't by lowering his blood pressure or providing opportunity for contemplation. "Look, I used to be a serious climber. When I came off a big wall like El Capitan, I felt quite relaxed and it also felt good to be alive. It didn't feel restorative at time, but it was. I behave differently for weeks after coming off a climb."

It makes sense that going into a totally different, novel environment, be it an ice cave or a Club Med, can be a great antidote for day-to-day stress or drudgery. That's the recovery piece. But what about the source of stress? Compared to our ancestors, there's no doubt that modern life does challenge us with unique attention loads, and most of us have not figured out how to thrive under them. Levitin writes: "The average American owns thousands of times more possessions than the average hunter-gatherer. In a real biological sense, we have more things to keep track of than our brains were designed to handle." The fact is, there's generally not a lot we can do about the stressor side of the equation.

And this, as Strayer explained to me, is part of our problem. "We are products of our evolutionary environment. We create artificial environments. Primates are good at being able to manipulate our environment and adapt, but that's not necessarily most consistent with the way we think." In other words, the world of office towers, traffic lanes and email isn't ideally suited to our brains' perceptual and cognitive systems. So what exactly are those systems? It's worthwhile taking a moment to lay them out, because they get to the crux of the nature-brain connection and the best ways to salvage it.

The way Strayer sees it, moving through any environment engages three main networks in the brain. There's the executive network, which includes the intellectual, task-focused prefrontal cortex and does most of that stimulus and behavioral inhibition. There's the spatial network, which orients us and does what it sounds like. Then there's the default network, which kicks in when the executive network flags. They are yin and yang, oil and water, working only in opposition. You can only engage one or the other at any point in time.

The default network is our free-ranging, day-dreaming, goal-setting, mind-wandering white noise that James so bemoaned for luring us from the real work to be done. But it is also the charismatic, elusive flower child of the brain. There's much discussion these days about whether the default network is profligate, undisciplined and troublemaking, or the very stuff that poetry and human nature is made of. When people are overly ruminative, depressed, self-involved and self-critical, the default network is blamed by psychologists. Yet it is also credited with producing empathy, creativity and heights of insight. Attention scientists worship at the altar of this network, because "it gives us our most human experiences, our deep aesthetic sense, our ability to do the deep things that are unique to us," as Atchley put it. That sounds exalted, but there's another important and more pragmatic reason they like it: it allows the executive office of the brain to rest, all the better to rebound at top performance.

One of the compelling theories about nature is that it acts like an advanced drug, a sort of smart pill that works selectively on the default network in the way new estrogen therapy makes bones stronger by targeting some estrogen receptors in the body but not others that might increase cancer risk. It would appear that when we have a positive nature experience, it engages what's good in the default network without allowing us to wallow too much in what's problematic. Studies show that when people walk in nature, they obsess over negative thoughts much less than when they walk in a city.

Although we can't always do much to turn off the barrage of stressors in our lives, we can try harder to get the restorative reprieves—from quick nature doses to longer ones—that give our thinking brains a chance to recover. In Utah, I was beginning to feel it.

Once I started thinking of the brain's oppositional parts, it was easy to watch the default network kick in on Hunter Creek. At first, I was all executive. Sunscreen? Check. Water bottle, bee sting meds, jalapeño potato chips? Check. Am I hungry? Of course, but I must wait until it becomes socially acceptable to eat. Do not think about the potato chips. Stop that. Chocolate nibble? Nope. I walked along, feeling the sand move beneath my boot. Tamarisk branches brushed against my leg, opening up to reveal small, brackish pools of water. The birds were singing; the flowers were outrageous. It was impossible not to notice them. I was beginning to become more sensory and less analytical, or what neuroscientists call bottom-up instead of top-down. The older parts of my brain were reasserting themselves over the chatty neocortex. It simply doesn't usually require intense concentration to walk across a landscape, one foot in front of the other, at the speed of human locomotion. This is a speed our brains naturally understand.

During lunch atop warm boulders near the creek, I pulled out my flower guide. We lumbered down to gather around a white blossom on a stalk. Turns out there were quite a few of these on the laminated card, and this one didn't quite fit. "I think it's a buckwheat," said someone. "No, look at the leaves. They're pointy."

"That's gotta be this one, a milkvetch," said Atchley, pointing to the card.

"Actually, it's a stinking milkvetch."

It was natural history by committee: educated guesses, disputes and confident pronouncements that turned out to be wrong. It was probably a lot like doing brain science.

THE IDEA OF nature as a kind of orchestral conductor of attentional resources isn't all that new. Remarkably, Frederick Law Olmsted wrote of exactly this phenomenon in 1865, arguing that viewing nature "employs the mind without fatigue and yet exercises it; tranquilizes it and yet enlivens it; and thus, through the influence of the mind over the body, gives the effect of refreshing rest and reinvigoration to the whole system." Slowly, slowly, academia started to catch up. Beginning in the early 1980s, Stephen and Rachel Kaplan at the University of Michigan noticed that psychological distress was often related to mental fatigue. They speculated that our constant daily treadmill of tasks was wearing out our frontal lobes. This part of the brain got exercised in premodern life too, but the difference is it also got more rest, said the Kaplans.

Before coming to Moab, I had spoken with Rachel Kaplan, who

works from her plant-filled university office in Ann Arbor. She and her husband are still revered within the world of environmental psychology, and together their mentorship has spawned dozens of leading researchers around the world whose work crops up across these pages. What leads to brain-resting? I had asked her. "Soft fascination," she'd said. That's what happens when you watch a sunset, or the rain. The most restorative landscapes, she said, are the ones that hit the sweet spot of being interesting but not too interesting. They should entice our attention but not demand it. The landscapes should also be compatible with our sense of aesthetics and offer up a little bit of mystery. You can find these conditions indoors if you're lucky, but they spring easily from natural environments.

The Kaplans called their hypothesis the Attention Restoration Theory, or ART. They tested it qualitatively at first, finding that their subjects expressed clearer thinking and less anxiety after viewing nature photographs or spending time outdoors. In 2008, Stephen Kaplan teamed up with one of his graduate students, Marc Berman, for more empirical testing. They found that short sessions of nature-image viewing (compared to pictures of urban setting) allowed subjects' brains to behave as if at least partly "recovered," specifically in measures of cognitive performance and executive attention. Rachel Kaplan thinks these effects will only get bigger as time in nature increases.

One of the Kaplans' early students was Roger Ulrich, the EEG researcher we met briefly in the last chapter. While the Kaplans promulgated the idea of attention restoration, Ulrich instead argued on behalf of the Stress-Reduction Theory, or SRT. It's worth pointing out the main difference between ART and SRT, and it's mostly a question of timing. Both propose that nature makes us happier and smarter. In the Kaplans' ART theory, the first stop is the brain's attention network. Nature scenes, like my walk up Hunter Creek, lulls us with soft fascination, helping to rest our top-down, direct-attention faculties. With that restoration, we become more relaxed, and then can perform thinking tasks better. SRT and Wilson's biophilia, on the other hand, posit that nature exposure can immediately lower our anxiety and stress levels, and then we can think more clearly and cheer up. Ulrich explained the intellectual split with the Kaplans to me: "After getting my Ph.D. our paths diverged with respect to conceptual thinking and research methods. Their work continued to evolve around cognition. Mine turned in the directions of emotional, physiological, and health-related effects of nature." Ulrich influenced the Japanese with their blood-pressure cuffs and mood scales, while the Kaplans' attention framework has generally held more sway with the Americans.

"How could we have possibly imagined where all this would go?" asked Kaplan, marveling at the long tail on the creature she and Stephen birthed. Both ART and SRT still leave plenty of room for investigation: What constitutes soft fascination? Through which sensory systems do we register the scenes that change our moods? How do you define nature and how quickly do these responses occur?

Here's Team Moab's overarching hypothesis: After days of wandering in a place like this, resting the executive branch and watching the clouds drift across an endless sky, good shit happens to your brain.

"After three days, there's just this feeling, ooh, something changes," said Paul Atchley.

Added Strayer: "We'd be foolish to ignore it. By the fourth day, you're more relaxed, you notice details. In the wilderness, there's a novelty effect for the first few days, you've got a new backpack on, there's all this equipment. But then the novelty wears off and that novelty was attracting your attention, so now your attention is not grabbed. There's a capacity to use other parts of your brain. It's like when Michael Jordan had the flu when the Bulls played the Utah Jazz. You can't write him off because he plays well like that. He scored thirty-eight points in a row. He was mindless." His executive network was not in the house. He performed better, flying on pure intuition. We've known for a long time that athletes and artists can easily access flow states; the idea that the rest of us can touch that zone through nature is tantalizing.

"Down with the frontal lobe!" said Atchley, bounding back down the trail after lunch, his hydration-pack tube trailing behind his neck. "Up with the cerebellum!"

LATER THAT NIGHT, Gazzaley mixed martinis by the rooftop fire pit. If Kramer is the senior member of Team Moab, Gazzaley is its boy wonder. At forty-six, his premature bright white hair belies his youthful face. It's so incongruous that people sometimes ask him if he dyes his hair.

"Dye it this color?" he pointed to his head, barking a laugh. Extroverted and optimistic, Gazzaley is refreshingly unapologetic about his affection for technology. He believes it is not our curse but very possibly our salvation. He employs his gadgets with ease and fluency, from his cameras to the brain-wave monitoring machines and 85-inch high definition screens in his multimillion-dollar laboratory at the University of California, San Francisco. There, he is designing and testing "neurological" video games built specifically to increase cognitive performance in adults. The games, he believes, can help prevent dementia, treat ADHD, and even make us all better multitaskers, and he has data to back it up. This is the world we live in. We might as well get better at it.

Still, as a nature photographer and adventurer, he is loving the desert. He had his vertical-panorama revelation yesterday, and he had another spark of insight today in Hunter Canyon. "I had such a rich experience of flow today," he told us around the fake campfire. "I was walking in the sandy canyon. Dave took off in front of me, and I found myself alone taking pictures of desert flowers. I made myself receptive to the stimuli around me. It was so bottom-up, moving through the environment and it was all fitting together. I usually have trouble not being top-down, but without trying to, I was picking up things that were beautiful and salient. I realized how natural and comfortable and smooth it felt to do photography. I'm always thinking about top-down versus bottom-up, and I usually present it as conflict, basically, over cognitive control, but the insight was as it relates to flow and it's that maybe it happens when these parts of the brain are in perfect balance. I hadn't felt it in years and it felt really good."

There was more, because his analytical top-down mode was in full force now. Gazzaley the neuroscientist was back. He had, essentially, experienced Kaplan's theory about attentional restoration. The Queens techie was drinking the Kaplan Kool-Aid, along with the martinis: "Nature is restorative because it frees up the top-down part of your brain in a way that allows it to recover. I don't think you have to be in nature for this to happen, but I think there's something special about nature. It's what makes it interesting. Nature has this not totally unique but more powerful ability to capture your attention in a different way. Evolutionarily, nature is a powerful bottom-up experience for us." He paused and then laughed. "Although a lot of people freak in nature. I've seen it countless times."

Ruth Ann Atchley piped up. "I was not restored while hiking the fins yesterday. I do not like heights."

Lisa Fournier apologized for the route.

Strayer: "There are always going to be individual differences." Here I couldn't help thinking of Woody Allen: "I love nature, I just don't want to get any of it on me."

Fournier was thinking. "Nature is pretty novel in lots of ways. You're immersed and enriched."

Dyre, the skeptic: "Maybe it's the active exploration that's important."

"Yes!" said Jason Watson, a young researcher and associate, another attention scientist who'd become captivated by the nature effect and whose shyness dissipated under the night's half-moon. "It's what Kaplan calls mystery." Watson told us about a recent study he'd done that largely confirmed Kaplan's mystery element. He and his colleagues showed a couple hundred subjects images of nature scenes, some with flat, predictable trails and some with winding or partly obscured scenery, the kind of images that compelled the viewers to want to peek around the corner. Even though the subjects saw the images very briefly, just a matter of seconds, they remembered the mysterious scenes better. In other words, there was something about mystery that improved cognitive recall.

Ruth Ann Atchley saw a good transition point. "Okay, I have one question: what kind of studies should we do now?"

"What I'd like to know more about is creativity. We can do cognitive tests, but we also need biomarkers," said Strayer.

Art Kramer had helped find a beautiful biomarker, the neural growth factor BDNF, which spritzes the brain like Miracle-Gro during exercise. Could nature exposure unleash some similar, visible molecule? Until recently, it's been hard to see inside the brain in real-world settings or under more sophisticated lab conditions. Some studies show a drop in hemoglobin levels (a proxy for blood and oxygen) in the prefrontal cortex during time in nature. It's still debatable where the blood is going instead. At least one MRI study (using photographs of nature) shows it's going to parts of the brain like the insula and the anterior cingulate that are associated with pleasure, empathy, and unconstrained thinking. By contrast, when those same subjects viewed urban pictures, more blood traveled to the amygdala, which registers fear and anxiety.

Strayer would like to know what a brain looks like as it's getting restored. Can you see it? Does it look different in the real world compared to in a lab that uses photographs? After some discussion,

Gazzaley proposed they use EEG—electroencephalography—to measure brain waves, specifically one called frontal midline theta, which his lab has found to be a reliable measure of executive-center engagement. If it quiets down in nature, that could be evidence of what he experienced on the trail: less top-down, and more bottom-up, less executive network, more default network. It would indicate a rest break for the frontal lobes.

"I love it!" Gazzaley said.

They discussed the complications: Strayer prefers field data and not lab data. He wanted people wearing the caps in real nature, not just looking at pictures of it in an air-conditioned room. But Kramer and Gazzaley prefer the controlled environment of the lab. Kramer would leave Moab with a plan to study whether creativity differed for people walking on a lab treadmill looking at virtual-reality city images versus nature images. I made a note to check back.

"It is messy, no doubt about it," said Strayer of working outside. "You can study this in the lab, but for the effects to be there, you have to be in nature. People said we couldn't measure the effects of driving and distraction in the real world, because there are so many variables, but we did it." Strayer would leave with several experiment ideas: a walking study in an arboretum measuring creativity, and another using the EEG on a group in the wilderness. This I would have to see.

Gazzaley had a plan for yet another study. Nature, he saw from his own Kaplanesque moment of "flow" out on the trail, could be useful. It could improve not the way we enjoy nature but the way we use technology. "My practical desire is to understand how to maximize our brains," he said. "If I'm going to build software to enhance cognition, what if I routinely inserted recovery periods in virtual nature? I'm a fitness buff. You have to rest between sets. Everyone knows you can't just blast your brain for hours with video games or you get diminishing returns. Are all breaks equal? I'm going to try nature."

The Atchleys, for their part, would also soon run an experiment to see if group problem-solving improved among workers outside versus workers inside.

I'd have to stay tuned. The trip had crystallized for me some critical questions to keep in mind as I moved ahead. If nature environments have the potential to change both our emotional brains and our cognitive brains, how would different doses of nature affect us? How much of the benefits of nature are really because of what's in nature versus simply leaving behind the bad stuff of cities and workplaces? And, based on what I would learn about our perceptual systems, how could we improve our normal lives back at home?

For science, I was learning, you have to be patient. But maybe you can draw a payoff like Gazzaley's pursuit of an American three-toed woodpecker in Rocky Mountain National Park. Before the moon set, he pulled up some of his photographs on his laptop and scrolled through them for us. The bird was coy, finally poking his spectacular black-and-white-striped head out of a hole in a tree. But Gazzaley was ready, camera in hand.

"I had to wait six hours for this fucker," he said.

Together and apart, the group would be looking at the puzzle of nature and the brain from many angles. As Paul Atchley put it at the end of the evening, no doubt inspired by the night sky, the beverages and a new laser focus in his attentional network, "It's many fingers pointed at the moon. If you look at all the different fingers, eventually you can see where the moon is even though every perspective is different. There won't be a single piece of evidence. Science doesn't work that way."

These and other emerging studies would make up the next frontier in understanding nature's role in optimizing human potential, many aided by brain imaging. With more clues about what makes our brains happy and keeps them running smoothly, that information can be fed into public policy decisions, urban planning and architectural design. The research has profound implications for schools, hospitals, prisons and public housing. Imagine bigger windows, more urban trees, mandated lie-on-the-grass sessions, minute-long birdsong breaks. Per Gazzaley's quest, it might even be possible to construct doses of nature so palatable and efficient that we hardly notice them. This approach, of course, is classically Western. Manipulate the environment. Feel nature without even trying.

As for me, I would be looking for a more East-meets-West approach. I would come close to finding it in Korea. That country has wrapped a pervasive wellness philosophy around the senses, particularly the sense of smell that builds on the work from Japan. It's a good place to start the next section, which looks at the immediate benefits of nearby nature.

PART TWO

NEARBY NATURE:
THE FIRST
FIVE MINUTES

The Smell of Survival

I can't begin to count how many times I was on some kind of a
trip with my parents and they woke me up at dawn because it was
mandatory that I watch the fucking sunrise.

—EUNY HONG, *THE BIRTH OF KOREAN COOL*

Park Hyun-Soo didn't look like a man on chemotherapy.
Forty-one and with a full head of black hair, he can hike the
socks off anyone, but he prefers to take his time. I met him
after a basic country lunch of eight kinds of kimchi and a plate of
neatly sliced homemade tofu. Eating the tofu was a little like biting
into air and earth at the same time, a barely solid cloud of unde-
manding goodness. The kimchi, on the other hand, had a flavor as
subtle as a firecracker. Each slice of cabbage, sesame leaf, radish
and mystery veggie had been rubbed and soaked in hot chiles, garlic
and anchovy paste. I went light on the kimchi but I'd eaten too much
tofu. If Korean food is all about balancing flavors, I was clearly lop-
sided, as Americans tend to be. We like the easy pale food. I felt the
need to walk briskly, but that wasn't about to happen.

First, there was tea. Not exactly a forest ranger, Park is more of a
ranger-slash-shaman. Remarkably, that is pretty much his official
job description. He is part of a new breed of Korean Forest Agency

employee known as a forest healing instructor. He'd actually gone to graduate school for this, passing rigorous entrance qualifications. He did not always aspire to this profession. He began his career as many South Koreans do, in a competitive corporate job—in his case, general manager of a hospital clinic in a city a few hours south of Seoul. But then, at age thirty-four, he received a diagnosis of chronic myeloid leukemia. He had a wife and three small children. He sought peace and recovery in the nearby woods, and it worked so well he decided to orient his entire life to the cypress trees. Here, in his mountain aerie, he stands at the forefront of South Korea's project to medicalize nature, beginning with its immediate sensory effects.

Park greeted my translator and me in the visitor center parking lot of Jangseong Healing Forest and ushered us inside. The building was brand-new, constructed of blond woods and redolent of the pleasant, slightly acrid smell of hinoki cypress with its robust notes of turpentine-meets-Christmas tree. Park apologized for the low table in the conference room, asking me if I'd be okay sitting cross-legged on the floor. Of course! I said. Not all Americans are stiff-legged blobs of hopelessness. We drank the tea, made from benzoin tree flowers harvested here in the summer. After twenty minutes I desperately had to shift position, and pined once again for the promised walk. He was telling us that between 2,000 and 3,000 visitors come through here every month, including three to four groups per day specifically geared to some kind of healing, from cancer patients to kids with allergies to prenatal groups and everything in between. Depending on the program, participants may do activities like guided meditation, woodcrafts and tea ceremonies. But the heart of it all is walking in the hinoki forest. Yes, please!

I creaked up from the table and wobbled into the physiology room. Like all the participants, I would capture a snapshot of my stress levels before and after the program, although for me, the agenda would just be a walk, a quick squirt of cypress mist and a few moments of deep breathing. That is because, as usual, I was too busy for full-on relaxation. I had a full schedule of forests and scientists to see on my week around South Korea. Today could be called the mini-jet-lag-and-tofu-recovery program. My translator, Sepial, was even more harried than I as she had to keep track of every exchange while still responding to emails and setting up visits for me later in the week. She's forty-four, with a teenaged son. She needed a little walk in the woods herself. "I don't usually exercise, Florence," she said, looking apprehensive.

We took our blood pressure and then inserted a finger for several minutes into a plastic clamp sensor that is supposed to measure our heart-rate variability. The idea was that the Korean Forest Agency will keep all these records and use them to assemble a large database for research. Individuals will be able to track their own data over time and across different forests and facilities. They should be able to tell if one walk in the woods per week is enough for them to maintain lower blood pressure, or if they better try adding more leaf-and-acorn collages to their regimen. The scope of all this was, true to Korean form, ambitious. In the same way Samsung outmaneuvered Apple and K-Pop intends to dominate Asia with American-derived pop music models, Korea is on a path to out-Japanese the Japanese in forest therapy trails and science. Here, forest bathing is called *salim yok*.

Although Jangseong is currently one of only three official healing forests in South Korea, thirty-four more are slated to appear in the next two years, meaning most major towns will have access to one. This forest, with its dominant cypress trees, is considered a jewel in the system. Finally, I was able to see it. We headed out to walk, first following a wide dirt road through the woods and then branching off into a narrow, well-maintained foot path. The trail skirts around 2,900-foot Chukryeong Mountain. We passed an interpretive sign claiming the woods have more oxygenated air than a city or a building, although I wondered if this isn't offset by the gain in altitude to thinner air.

Park wore what looked to be comfy Mao-style pajamas, with a round wooden nameplate attached to his chest. He moved gracefully along while recounting the history of this ground. Like much of Korea after World War II, these mountain flanks were once completely denuded of trees. First the Japanese, who occupied Korea starting in 1910, cut the forests for timber. After the war, people scavenged whatever was left for heating fuel. Times were desperate. At $100 per capita, South Korea then had a GDP lower than that of Ghana. One-third of Koreans were homeless. Without trees to anchor the mountain in place, the mud slid and the streams choked with silt. Replanting began in earnest in the 1960s. The Japanese hinoki cypress was a favorite for its fast growth and uncanny ability to ward off pests. Jangseong is now 88 percent hinoki, and the trees are fully grown.

What makes the tree so unappetizing to insects has vaulted it to the heart of the Korean Forest Agency. It smells great. Walking through Jangseong is like moving through a picturesque vat of VapoRub. Whether or not these woods noticeably increase our oxygen supply, it feels like they do, clearing the sinuses and infusing every cell with an essence of the forest, something healthful and invigorating. Robert Louis Stevenson has a line about "that quality of air, that emanation from old trees that so wonderfully changes and renews a weary spirit." He had a good nose. So did D. H. Lawrence, who wrote (or rather overwrote): "The piny sweetness is rousing and defiant . . . keen with aeons of sharpness. . . . I am conscious that it helps to change me, vitally. I am even conscious that shivers of energy cross my living plasm, from the tree, and I become a degree more like unto the tree, more bristling and turpentiney. . . ."

Clearly, cypress trees and the love for them are not unique to

Asia. They are prized the world over for their rot-resistant wood, warm tones and pleasing scent. In ancient Egypt the tree was used for mummy cases. Cypress wood was even thought to outlast brass, and so it served as a palimpsest for Plato's code of laws. With its rich amber bark and soaring greenery, Jangseong felt comforting, almost congregational. While I'd walked in forests in Japan, the ones I saw bore a mix of hardwoods, cypress, other native evergreens. Jangseong, though, is practically a mono crop.

In what I understood might be the Asian conception of nature, compromise would do just fine. It doesn't have to fulfill an Emersonian purity in order to be considered sacred. I asked Park about wildlife, and he admitted there is not much here in the way of large mammals. Most have been hunted or squeezed by poor habitat into the surprisingly biologically rich Demilitarized Zone between North Korea and South Korea. People have been locked out of that 160-mile long, 2.5-mile wide buffer for decades, making it a prime candidate for an international peace park, if only North and South could agree on anything.

What these woods lack in biodiversity they make up in sensory delight and, increasingly, human medical use. "There are two and a half million individual trees here," said Park. A subtle mist rose from them, made of the very aerosols we were smelling. Atmospherically, these serve a cloud-seeding function, helping forests regulate their moisture levels. But Park, healing instructor that he is, holds a strictly medical appreciation. "The phytoncides are antibacterial," he said. Citing the Japanese research of Miyazaki, he continued as though he's recited it many times before: "They reduce stress fifty-three percent and lower blood pressure five to seven percent. The soil is also good for healing. It is antiviral and the geosmin is good for cancer." Geosmin, I learned, causes the funky-great smell of earth after a rain. Like many of the phytoncides, it is a turpene, a family of aromatic hydrocarbons and a major component of natural resin (incidentally, turpenes are also a big ingredient of hops, giving dark beer its rich flavor and aroma).

Geosmin comes from soil organisms, particularly the streptomyces bacteria that are key to so many antibiotics. According to the Royal Society of Chemistry, we are alert to this rich smell in incredibly small quantities. We can detect the equivalent of seven drops of geosmin in a swimming pool. This sensitivity likely reflects an important evolutionary adaptation because it tipped our thirsty ancestors off to sources of water. That may also explain why its presence helps put us at ease. Camels probably get off on it even more than we do. Keith Chater, the Norwich scientist who sequenced the genome of *Streptomyces coelicolor* in 2007, believes camels can smell geosmin in oases miles away. In return for their helpful homing service, some spores of the bacterium then hitch a camel ride to the next watering hole. Geosmin is the smell of survival.

It's no surprise by now that Korea and Japan lead the world in the science of forest smells. There's the Natural Killer cell work of Japan's Qing Li, and also that of a young psychologist there named Yuko Tsunetsugo. A senior researcher with the Department of Wood Engineering at the Forestry and Forest Products Research Institute, Tsunetsugo misted fifty-two infants with the major components of hinoki: pinene and limonene. The pinene instantly lowered their heart rates four points, while the limonene and the control did not make a difference.

When I'd been in Japan at the Nippon Medical School lab of Qinq Li (the man who put subjects in hotel rooms for three nights with hinoki oil misting around them), he'd given me a demonstration of the immediate effects of the stuff. I'd put my arm in a blood-pressure cuff. Then he unscrewed the cap off the forest elixir. "This is very toxic!" he'd giggled. "It's very good but very toxic." When I inhaled, the oil gave off a nice, pitchy, sharp scent. We put the cap back on and read my blood pressure again. It dropped twelve points.

I'd looked at Li, who nodded delightedly. "This is a very big effect, bigger than people get with pharmaceuticals!"

Meanwhile, here at the government-funded Korea Forest Research Institute, scientists distill essential oils and study them for effects on allergies and their ability to kill staph bacteria. Among the things they've found are that coniferous essential oils fight atopic skin diseases (when applied to the skin in low concentrations), mitigate stress by lowering levels of cortisol (when inhaled), and reduce symptoms of asthma (ditto). The major components of hinoki oil are camphor, turpenes, pinenes and humulenes, limonenes and sabinenes, depending on the season and the part of the tree sampled. The sabinenes seem particularly helpful for treating asthma, the terrines for fighting bacterial infections and stress.

I may not have been actively nursing any infections, but after a few minutes of walking I felt more awake than I had all day. We stopped where a wooden boardwalk crosses a small wetland lined with dogwoods and connects two drier parts of the trail. Park pointed out a citronella plant and a Japanese cedar, also prized for anti-infective properties. He asked us to close our eyes and take deep breaths. Then he led us in some gentle stretches. Sepial stashed her notebook into the recesses of her trench coat. We raised our arms over our heads, then down, then back up, all while breathing slowly. The birds chirped. The wind blew gently through the high branches, and the sun mixed with the cool autumn air. He told us to look at the still pond of water just beyond the trail. "Look through the lake, watch the reflections of the trees. This is good for the brain to see. Pretend this is your mind. Take deep breaths. The trees you see there could be real, or they could be fake, just reflections. This is like your mind. When a depressed person sees depression, it could be an illusion. It's not really there. You can separate the emotion from the mind."

Maybe it was the translation, but things seemed to be bleeding out of the realm of quantifiable science and into a squigglier place.

Was the mysticism biasing the science and making it suspect, or was it more like a portal allowing the scientists a point of entry where Westerners don't always feel comfortable? Or a little of both? I wasn't sure.

FOR THREE YEARS, Park had walked mindfully in these woods every single day. "I'm one hundred percent sure it is helping me," said the ranger, who is in remission. "When I was first diagnosed, I had all kinds of fear and anxiety. I am happy now. I have zero percent anxiety. People learn from nature that they can heal. Now it is my duty to be a bridge between nature and people." He said he's grateful to the leukemia for redirecting his life. It's hard to say, though, what's really helping Park and the many who flock to these places. Is it the exercise? Park wears a bracelet that measures his steps. He takes 15,000 a day, about 6 miles. He also believes the forest heals him, and the power of belief is hard to overestimate.

It also may be contagious. Park is a compelling teacher who wants to help other people turn away from stress and toward something more meaningful than the punishing grind of work and study. He doesn't force his kids to attend the pervasive after-school schools—called hagwons—that so many kids slouch off to, forgoing sports, play and just goofing off. His oldest son now attends a "timber school" for high school where he learns about forest management.

Park told me he thinks Korea has entered "Peak Stress." It's an interesting idea. Flying out of poverty and through a series of dictatorships to become one of the wealthiest democracies on the planet, the nation now boasts the fourteenth-strongest economy in the world. An incredible 98 percent of South Koreans graduate from junior college or university, the highest rate in the world. But the meteoric success has come at a great cost. South Koreans work 2,193 hours per year on average, the highest figure in the OECD. More than 70 percent report their jobs make them depressed, according to a survey by one of the country's biggest employers, Samsung.

And the problems aren't confined to the workforce. Ninety-six percent of high school students reportedly do not get enough sleep. A 2011 survey found 87.9 percent of them feeling stress "in the past week." Teenagers in Japan, China and the United States report half that level. South Koreans are, according to researchers at Yonsei University, the unhappiest students in any industrialized nation. In a country where mental illness is highly stigmatized, South Koreans have the highest suicide rate in the world.

But now that they've achieved some measure of security and material success, some are actively seeking a happier existence. South Koreans are buying into the booming spa and cosmetics cultures, and, increasingly, yearning for the mystical mountains and forests of the deep Korean past. Since it arrived here in the fourth century, Buddhism blended nicely with the peninsula's ancient animistic shamanism, the idea that natural objects have a spirit. In Korea, one of the most powerful spirits is the *sanshin*, the mountain spirit. Trees, too, have long been venerated as guardians of people and villages.

By the fourteenth century, though, Korean rulers would find in China-originated Confucianism—with its teachings of regimented status, societal obligations and an uncompromising work ethic—a politically convenient philosophy for growing a nation state. There now exists an uneasy and unequal détente between opposites: a technology-touting, competitive and hierarchical system on the one hand and the nature-affiliated spirits-are-everywhere firmament on the other.

Euny Hong, in her irreverent cultural history of Korea, *The Birth of Korean Cool*, explains an ancient proverb, *"shin to bul ee,"* which means "body and soil are one." Not soul, but soil. "It's a concept that predates Confucianism or any official organized belief," she writes, "which is why this idea seems incongruous with what Seoul looks like today—jam-packed skyscrapers with very little open space."

While most Koreans would be uncomfortable with the idea of psychotherapy, they do nonetheless place great authority on traditional shamanlike healers, called *musok-in*. By some estimates, up to 80 percent of Koreans loosely adhere to shamanism in some form, often while also identifying as Christians, Buddhists or atheists.

What it means today is that the forest trails are starting to fill up with pale, urban weekend refugees, not so unlike Sepial and me. After about an hour and half of leisurely walking, we circled back to the visitors center. We gamely stuck our extremities back into the machines for a quick physiology check. I clocked a nice little drop in my blood pressure, from 111 over 73 to 107 over 61. So far, chalk one up for Nature. But Sepial's blood pressure was a few points higher, and my heart-rate variability data didn't show much improvement after the 90-minute walk. Park sat down with us to go over the charts, which were in Korean, with confounding splashes of dots strewn across an axis. Looking at Sepial's data, Park told her that because she wasn't used to exercise, the walk had actually stressed her out physiologically. "You need to exercise more," he said. It seemed a logical prescription. Don't health-care practitioners always say that?

As for me, Park said that while my overall stress levels seem healthy, my chart indicated that the balance between my sympathetic nervous system and my parasympathetic nervous system is out of whack. I know how to amp my system up with exercise and activity, but I could use more practice damping it down. In other words, Sepial and I appeared to be opposites. "Meditation could be good for you," he said. In more bad news, the HRV machine mysteriously gave a read on how thick my blood vessels are. Mine were showing some signs of thickening, and any time the word "thickening" applies to you, it's not auspicious. Vessels naturally thicken with age, getting stiffer and less flexible. They have a harder time delivering oxygen where it needs to go and making micro adjustments to the nervous system. "You must control your food and diet," he said. Okay, then: more kimchi for me.

WHAT HAPPENS IF you take someone with a fairly radical notion of happiness and set him loose to make national policy? The answer might look like Bhutan, where the king and his retired-king father ride bicycles up and down mountains with shit-eating grins on their faces and encourage the populace to do the same. Or it might look like Singapore, where the late Lee Kuan Yew, the prime minister for twenty-five years, ordered free schools, decent housing and the planting of over a million trees. Increasingly, it might look like South Korea. The man with the grin on his face is an influential academic named Shin Won-Sop.

To understand just how committed Korea is to better-health-through-forests, I paid a visit to the headquarters of the Korean Forest Agency in the new industrial city of Deajun. There I was pleased to find my old *shinrin yoku* contact Juyoung Lee, who'd been hired away from his post in Japan to conduct research for South Korea. Lee now works for the agency's human welfare division. It's remarkable that any forest agency even has a "human welfare" division. It wasn't so long ago that the main job of forest agencies the world over was simply to facilitate cutting down forests. When I first met Lee two years earlier, he was swatting mosquitoes and suctioning sensors off my forehead on a Japanese mountainside. Now he wore a stylish suit in a modern high-rise filled with pink cubicles. (Not sure what the significance of the pink was, but I can't resist reporting that the city of Seoul recently spent $100 million painting special parking spaces pink for women. They are supposed to make women happy, but they are also longer and wider, leading many not to feel happy but to feel insulted by the implied dig on their driving ability.)

Lee escorted me through the maze of pink to the spacious outer office of Dr. Shin, who is the minister of the Korean Forest Agency. Shin greeted me with a handshake and a delicate cup of tea. He is boyish and buoyant, as if he can't quite believe his good fortune to land the corner office. He did not rise to the top of the agency by the usual route in timber management, but rather because of his psychology research on such topics as "the influence of inter-action with the forest on cognitive function" and "the influence of forest experience on self-actualization." For that paper, which he published while based at the University of Toronto, he studied how participants changed after a five-week wilderness course sponsored by the National Outdoor Leadership School and found the results inspiring. He'd been influenced by Stephen and Rachel Kaplan's work at the University of Michigan. Shin became a professor of "social forestry" at Chungbuk National University, which offers the world's only degree program in forest healing. In the early days of research, "we discussed a lot of the issues for how we can objectively measure the benefits and what are the best bio-markers," he said.

Apparently, the effort paid off. Shin's ascendancy and the country's new programs reflect just how seriously South Korea takes the emerging evidence on nature and health. The goal of the current National Forest Plan is "to realize a green welfare state, where the entire nation enjoys well-being." As Shin pointed out, happiness is now part of the national index. And the results of this campaign are evident: visits to the country's forests increased from 9.4 million in 2010 to 12.7 million in 2013, or one-sixth of the country's population (around the same time, visits to national forests in the U.S. dropped by 25 percent). The agency now offers everything from prenatal classes in the woods to forest kindergartens to forest burial options. It's a cradle-to-grave operation. There is even a "Happy Train" that delivers school bullies to a national forest for two days so they can learn to be nicer. To unwind in the United States, men in groups might hunt and drink Jack Daniels. Here they do downward dog and make floral collages. Earlier in the week at a forest named Saneum, I'd come upon a forest-healing program for firefighters with PTSD, where the men were practicing partner yoga in the woods and massaging lavender oil into each other's forearms.

The data on the healing power of the forests kept rolling in. Among the things the Korean researchers were finding: immune-boosting killer T cells of women with breast cancer increased after a two-week forest visit and stayed elevated for fourteen days; people who exercised in nature (as opposed to the city) achieved better fitness and were more likely to keep exercising; and unmarried pregnant woman in the forest prenatal classes significantly reduced their symptoms of depression and anxiety.

What's needed now, Shin told me, is better data on individual diseases and on the specific nature qualities that really deliver. "What are the main factors in the forest that are most responsible for the physiological benefits, and what types of forests are more effective?" he asks. "And the other thing, how do we make the people more interested? And discussing how that forest benefit can be applied in the medical field and the insurance field." The agency estimates that forest healing reduces medical costs, creates new jobs and benefits local economies.

In addition to designating dozens of official healing forests and constructing facilities there, the Forest Agency is building an ambitious $100 million forest healing complex adjacent to the country's iconic Sobaeksan National Park, complete with aquatic center, addiction treatment center, "barefoot garden," herb garden, open-air decks, suspension bridge and 50 kilometers of trails. It's hard not to think of this as Disney meets summer camp. Because make no mistake: as much as Koreans may yearn for meaning, they are pragmatists. The nature renaissance here is largely about consumerism, albeit a medical consumerism. The forest developments are public-private partnerships, where real estate and resort investments will generate profits, where shops will sell phytoceuticals (hinoki bath oil, anyone?) and where people will be able to return to their schools and offices more productive than when they left.

I glimpsed this hybrid future at a resort called Healience. Upon arriving at the bucolic setting near the Saneum Forest, I was handed a purple jumpsuit to wear during my stay, part Miraval, part Sing Sing. I joined others wearing these suits as we scrambled over barefoot forest-walking trails, waited for massages and bused our cafeteria trays. The lobby shop was a shrine to hinoki, selling atomizing humidifiers and artfully packaged glycerine soaps. I ended up with a tube of phytoncide toothpaste. It tasted like gnashing a holiday wreath between your molars. That's not what gave me pause about putting it in my mouth. I was having a hard time getting past the fact that phytoncide is basically pesticide. There's nothing coy about the name. "Cide" means "kill." I pictured ants crawling up the trees and dying in twisted, tortured poses while sending farewell signals to their loved ones. At the very least it seems like the place could benefit from some rebranding. Do we really want to brush with the stuff and hike on "phytoncide trails?" I was also, to be honest, skeptical of the whole aromatherapy thing because its primary adherents, at least in the United States, also lean toward crystal worship and misshapen footwear.

But the real story with these compounds is both more complicated and more interesting. In the quest to find out what exactly it is about nature that meshes with our minds, smells emerge as an undersung but powerful component. Visuals tend to get all the acclaim, but as Proust knew, nothing hits the brain's emotional neurons more powerfully than odor. Scents immediately enter the primal brain, where the amygdala is waiting to command a fight-or-flight response. The emotional amygdala is highly wired to the hippocampus, where memories are stored. A keen sense of smell was critical as we sought food and water in scarce environments.

Astonishingly, the human nose can detect 1 trillion odors, including many we don't even realize we are detecting. It's well known that women living together in dorm rooms are able to synchronize their menstrual cycles; the reason is they are nasally detecting each other's pheromones. Women may have a keener sense of smell than men, and it sharpens during pregnancy, when they must be alert to subtle hazards. Diane Ackerman writes in *A Natural History of the Senses* that mothers can identify their babies by scent alone, but fathers can't. My sense of smell is my sharpest sense, for better or worse. My nose detects hazards before my husband's, such as something burning that is not supposed to be burning, and it gets my heart beating very fast, a classic fear response.

We've all heard that horses and dogs can smell fear, but it turns out that humans can too. To prove this, researchers collected undershirts of men who went skydiving for the first time. They then presented study subjects with either those shirts or ones worn by men who did nothing scary. The researchers measured elevated stress hormones only in the subjects who smelled the skydiver sweat. They smelled the terror and then caught it too. Fear detection is a handy skill in a social animal.

Sadly, though, our brilliant sense of smell may be on the wane. Svante Pääbo is the Swedish paleogeneticist famous for sequencing the genome of Neanderthals and discovering that they interbred with early Asiatic humans (the result: all modern humans, except Africans). From genetic evidence, he posits we are drastically losing our sense of smell. We have a thousand genes involved in nasal reception, but over half of them have become inactivated due to mutations. In wild apes, only around 30 percent of the smell genes are dysfunctional. Presumably, the mutations persist in humans because losing some smell ability no longer affects our survival.

We no longer use our noses to find food, except perhaps Cinnabons in the airport. In fact, we would rather not experience many of the smells of city living. We refrigerate our food, but we don't refrigerate our garbage. Once proud, this superpower is devolving.

Certainly, we are not the sensory animals we used to be, and neither are the animals we've domesticated. Wolves outperform dogs in tests of general intelligence. Domestic cats differ from wild cats in some interesting ways having to do with skull size and foraging smarts. Which raises the provocative question: what about us? Are we self-domesticating? Of course, argues Harvard primatologist Richard Wrangham, who makes a particular case for humans becoming less aggressive as we've evolved into larger social groups. Our brain size and musculature peaked during the last ice age. Our teeth have gotten smaller, our long-distance vision worse. Since we settled down in farming communities around 10,000 years ago, we've grown weaker, and no doubt in some ways, dumber. The fast-firing sensory neurons we needed to stay alive in dynamic wild environments have, shall we say, relaxed. Of course we've gotten good at some things, like negotiating traffic circles and thumb-eye coordination for text-messaging. Scientists have shown that the hippocampi of London cabdrivers grow as they learn to map the city. Our individual brains are adapting to handle modern life, even from one year to the next, but that reflects flexibility, not evolution. In the mismatch between our current lives and our current brains, the primary victim is our paleolithic nervous system. No wonder, then, that when something smells really great we get happy. It's as though we've momentarily stepped through the wardrobe.

SMELLS HOLD POWER over us because the nose is a direct pathway to the brain. This is why some drugs are administered nasally. Molecules of a certain size that enter through the nose bypass the blood brain barrier and march right into the gray matter. While this shortcut is convenient for pharmaceutical companies, it's less help-ful in a world filled with pollution. Scientists have known for a long time that particulate matter from sources like diesel shortens life spans by causing cardiovascular and pulmonary problems. Black carbon—the tiny particles spewed out in exhaust and other combus-tion reactions like fires and cookstoves—are blamed for 2.1 million premature deaths annually around the world. Scientists have long considered the lungs as a primary target of pollution. Only recently have they come to realize the role of the nose as thruway to the brain; the nefarious extent of the nose-brain connection was only illumi-nated in 2003, when researchers in smog-choked Mexico City found weird brain lesions on stray dogs.

This is unnerving, because particulate pollution is all around us. It's very likely a strong factor in why going to the woods makes us feel better and more cognitively nimble. In the humid microcli-mates created by urban forests, leaves soak up particulate pollution. Beneath the trees, organic carbon in the soil can bind to airborne pollutants, and it also helps clean surface water in storms. A 2014 study estimated that trees in the United States remove 17.4 million tons of air pollution per year, providing 6.8 billion dollars in human health benefits.

I was curious about how the dynamics were playing out in my neighborhood. Before I went to Korea, I borrowed a portable aetholometer from Columbia University's Lamont-Doherty Earth Observatory. The device comes from the Greek word meaning "to blacken with soot." Velcroed into a twill vest pocket, it sent up a spindly arm sensor that poked out of my collar like a playful pet monkey. I wore it around D.C. for three days of my normal routine of working, walking and driving. Columbia's Steve Chillrud, codirec-tor of the Observatory's Exposure Assessment Facility Core, helped me collate the data to a real-time GPS tracker in my phone and analyze the results. Not surprisingly, I measured high readings of 6,000 nanograms per cubic meter while driving on I-495, the Capital Beltway, even during off-peak hours. More shocking, though, I recorded equally high values in my kids' school parking lots, where cars and buses idle waiting to pick up students gathered outside. Nineteen percent of Americans live near "high-volume" roads, and most cities don't monitor these corridors for air quality.

Regardless of your income, the closer you live to these roads, the higher your risk of autism, stroke and cognitive decline in aging, although the exact reasons haven't been teased out. Many scientists suspect it has something to do with fine particles causing tissue inflammation and altering gene expression in the brain's immune cells. "I hold my breath when I'm behind a diesel bus," said Michelle Block, a neurobiologist who studies pollution's effects on microglial cells at Virginia Commonwealth University. It's all another reason to spend time in the woods.

It makes sense that if some nasally routed molecules are bad for the brain, others might be good. We've known for millennia that smells can influence our moods, behaviors and health. Aromatherapy, or using fragrance specifically to help heal the sick, dates back to ancient Egypt. Cleopatra, that clever girl, reportedly used rose petals to lure Marc Antony to her bed. On a less legendary scale, retail stores and consumer product manufacturers know how to exploit the nose-brain connection. In the words of the academics who study such things, pleasant smells trigger "approach behavior." If a store smells good, we'll walk in and linger. In one study, participants cleaned their lunch area more assiduously if they smelled citrus. Even Windex changes our behavior. People assigned to a room sprayed with the pungent cleaner expressed a greater willingness to volunteer and donate money to a cause than participants in a neutral-smelling room. The hypothesis is that the smell of "cleanliness" makes us aspirational. Who knew: Windex is the smell of virtue.

When we say we can smell spring, we are really smelling tree aerosols. As the air temperature heats up, so do the biochemical reactions within the wood and leaves. Evergreen forests smell strongest in midsummer, which is also when pests are busiest. The so-called "pinosylvin" in pine trees and the terpinoids of cypress trees both stimulate respiration and act as mild sedatives, relaxing us.

Although aromatherapy is the most popular alternative treatment for anxiety worldwide, it hasn't been well studied in large, clinical trials. A review of the literature in 2011 found that while most studies showed beneficial effects, it was hard to tease out the power of the placebo effect in most of them. Nonetheless, the authors concluded it's "a safe and pleasant intervention." Since then, a large study found that 80 percent of cancer patients in the National Health Service of the U.K. reported significantly less anxiety while using "aromasticks." That's bigger than just a placebo effect, but the authors didn't know how the smells might be working. Other studies have reported that scents like lavender and rosemary cause both drops in subjects' cortisol levels and increased blood velocity to the heart (a good thing).

If you believe something can make you feel better, it sometimes does. The imagination is a powerful healer. Moreover, what if it's not necessarily nature that's helping us, but the absence of something else? Walking around sniffing the fresh hinoki forest, I had to wonder if some of the benefits attributed to these mystical woods are the simple result of not being in the city. If air pollution is so bad for us, getting out of town, even if it means sitting inside an aluminum box on a rural parking lot, might look pretty beneficial by comparison. Regardless of whether people know exactly how polluted their neighborhoods are, their psyches seem to know. In one survey of 400 Londoners, "life satisfaction" fell significantly—half a point on an 11-point scale—for each additional 10 milligrams per square meter of nitrogen dioxide pollution.

If less pollution makes us feel better, the same could be said of a reduction of noise, crowds, unwelcome distractions and, sometimes, technology. The latter is a big deal in Korea, the most wired country in the world. More than 90 percent of homes here have high-speed Internet access. As of 2013, the country had the fastest download speeds in the world, 40 percent faster than the number-two country, Japan, and six times faster than the world average. Video gaming is so big that it's a spectator sport, filling huge stadiums with fans watching sallow contenders push buttons on consoles.

In 2010 a young South Korean man collapsed and died after playing fifty straight hours of StarCraft, prompting the government to ban some games between midnight and 6 A.M. for anyone under sixteen. According to the National Information Society Agency, 8 percent of Koreans under age forty suffer from gaming addiction, with the figure rising to 14 percent for kids between the ages of nine and twelve. The government earmarked billions of won for counseling and education about the dangers of too much time on screens. These include poor grades, compromised sleep and family strife. Adults, meanwhile, evince slightly different symptoms. A survey of 500 office workers claimed their cellphones caused slouchy posture (32.7 percent), vision deterioration (32.5 percent) and finger pain (18.8 percent). The term "addiction" is controversial, but there are questionnaires to help identify distressing signs. Keeling over dead is a tip-off.

Perhaps it's inevitable that digital detox would find its way into the country's parks and forests. Nobody is happier to see it there than Kim Jooyoun. Like Park, she is one of the new healing instructors trained by the Korean Forest Agency. A mother herself, she understands the pressures on young Koreans and their striving families. Some years back, when her own daughter was fourteen, Kim found her literally pulling her hair out from stress. "Ever since then," she told me, "the child comes first." On Saturdays, Kim teaches a digital detox program for preteens in one of Seoul's big parks, Bukhansan. I visited on a glorious fall day, when hundreds of Koreans in smart outdoor attire moved like ants up the park's hilly trails. By the time I got there, seven boys were lying still like lizards on turquoise yoga mats in a relatively secluded grove. Kim was having them listen to the sounds of nature.

"If you want to play games better, you need to let your eyes rest," she told them. The boys' mothers hung about. This was week two of the free ten-week program, and they'd signed them up through the City of Seoul, having attested to their sons' obsessive behavior either playing video games like League of Legends or texting on their smartphones. I wondered why ten-year-olds even had smartphones, but that horse was clearly out of the barn.

I could see that Kim's forest program was as much for the stressed-out mothers as it was for the boys. The session included a clever mix of games, sensory interludes and trust exercises. Kim arranged everyone in a circle, each person holding a shoulder-high twig. Then she gave a command and each person lunged to the spot of the person next to him in time to catch their neighbor's twig before it fell. Then they switched direction. They made the circle bigger and the lunges faster. The boys, who had looked bored when it began, were soon laughing with their moms and stumbling into them. Next, Kim asked the mothers to put on blindfolds and allow their sons to lead them.

"I'm going to give you a chance to care for your mother since she's always caring for you," she explained to the boys. "The course where you will take her is not safe. There are lots of rocks and sticks." They walked carefully around for a while and then they switched places, the blinded sons alongside or just in front of their mothers. "Usually parents drag kids around with their intentions," said Kim to the moms. "The one who follows has no power at all, even though intentions are good. Don't talk too much and relax. If there is a tree in front, kids can sense it, so don't worry too much, and let the kids lead. Give them some space."

After that, Kim and her assistant led the boys on a slippery hike up a riverbed, one that would challenge them and not patronize them, she told me. It's not easy to compete with multi-player gaming, but she had the boys' full attention. The moms brought up the rear, stopping frequently for selfies. If the intention was to demonize smartphone use, they weren't exactly modeling good behavior. I learned, though, that tech abstention isn't the goal, any more than a dietary cleanse leads to anorexia. Unplugging isn't realistic, and seeing the Korean kids made me understand this in a new way. For many of these kids, gaming is the only play they get, and certainly the only play unsupervised by adults.

"They're not allowed to play outside at school," one of the moms told me. While there are spectacular parks in Seoul, they tend to be few and far between. Playgrounds are often covered in asphalt, small and claustrophobic. And the kids go to study programs after school, leaving little time for sports. They have it worse than their American counterparts, but I had to acknowledge that many of our kids, ever losing recess, unstructured play and time without adults, are not that much better off. No wonder they're meeting up in a galaxy far, far away.

Kim wants to help these families find a respectful balance of power between parent and child, an equilibrium between technology and human interaction, and healthier outlets for preteen anxiety, energy and aggression. She believes time outside can offer this. "In nature, they have to use all their muscles and senses. They develop body sense. They get scared but they develop self-confidence. They develop more ability to solve problems themselves."

The science backs her up. Two South Korean studies looked at eleven- and twelve-year-olds who qualified as borderline technology addicts. After trips to the forest of two days each, researchers found both lowered cortisol levels and significant improvements in measures of self-esteem, and the benefits lasted for two weeks. Time in the forest also led them to report feeling happier, less anxious and more optimistic about their futures, according to the lead study author, Park Bum-Jin, a professor at the Lab of Forest Environment and Human Health at Chungnam National University. A couple of days after Kim's program, I met him for green tea in the Seoul offices of the Korea Forest Foundation.

"Kids with higher self-esteem are less likely to get addicted," he told me. Based on this work, he recommends that preteens get out in nature for a half day or so every two weeks. "The philosophy of this research is simple," he explained. For these kids, "time spent in forest is not more interesting than video games, like fruit is not more delicious than junk food. We cannot make them stop playing games. As we get older, we have a tipping point in judgment that we need more fruits than junk food. As far as some time in forest, they can't play games during that time. As long as playing in forest is just fun itself, it can make that tipping point come earlier."

Park applauds the national plan that shepherds citizens into the forests through work and school programs. Koreans have been so intensively urban for long enough now—two or three generations—that they don't necessarily know what to do with themselves in the woods. In this Confucian culture of master and student, it makes sense to use rangers, guides and demarcated spaces—this hillside is for healing! This one is for plain old recreation! Camp on this platform here! Park pointed out that many Koreans have no hankering whatsoever to get back to the land, so it's especially important to catch kids early enough that they learn a sense of ease in nature. Interestingly, E. O. Wilson believes that the best window for the conditioned learning of biophilia is before adolescence.

The forest campaign can't come a moment too soon, Park said. He fears a loss of transmission from one generation to the next.

"Children and the younger generation don't really have experience in nature; so many of them think of the forest as dirty or scary. If we don't change their mind-set now, there will be no chance." Park himself, now in his early forties, grew up in the city with little time outdoors. Because of what he's learned, he takes his two kids hiking regularly. It's their vegetable, and they're dutifully consuming it.

Nature, for Park, is in some ways a negative space, a refuge from ills. It is the anticity, even when it's within a city. "Cities are a human zoo and I think schools are a human zoo too," Park continued. "We cannot give up those systems, city and schools. The forest is the only exit we have for those humans who live in the human zoo."

If the Koreans can learn to love nature, maybe anyone can.

Birdbrain

Most people never listen.

—ERNEST HEMINGWAY

Over the summer, I tried to find a patch of quiet. I spent some time wearing a portable EEG device on my head in different settings, trying to get a sense of which kind of places put me in the holy grail of brain states, the "calm alert" zone prized by Zen masters, surfers and poets. I was after alpha waves. When electricity in the alpha wavelength dominates parts of the brain, it's a sign that you are not hassled by small distractions, problem-solving or, my peeve, meal planning. Parenting—any kind of caretaking—is a procession of small, endless decisions. Too often, I assume the executive function for the whole family, and I can almost hear my mind stomping out any rogue alpha waves. It's the sound of brain fry.

Daily aggravations aside, environmental noise deters alphas because we have to either pay attention to the intrusion or actively resist paying attention to it, and that's work too. I couldn't quite hit the alpha zone walking in the city parks near my house, and I

couldn't even attain it on a leafy, rural road in Maine either, probably thanks to nearby construction noise, which ended up pissing me
off. When my brain waves were later read by the interpreting software, it fired back this message: "This indicates that in this state
you were actively processing information and, perhaps, that you
should relax more often!"

Even the software was yelling at me. I wanted to yell back, but
this would be a mistake. There are no alpha waves when you're mad.

And the maddening truth is, the world is getting louder.

Can you hear it? "Noise" is unwanted sound, and levels from
human activities have been doubling about every thirty years, faster
than population growth. Traffic on roads in the United States tripled
between 1970 and 2007. According to the U.S. National Park Service, 83 percent of the land in the lower forty-eight states sits within
3,500 feet of a road, close enough to hear vehicles. For planes, the
figures are even more dramatic: The number of passenger flights
has increased 25 percent since just 2002, and 30,000 commercial
aircraft fly overhead per day. In 2012, the Federal Aviation Administration predicted an astounding 90 percent increase in air traffic
over the next twenty years. Human activities in general increase
background noise levels by about 30 decibels. The official word for
the human-made soundscape is the anthrophone.

Stats like those above dismayed Gordon Hempton, a sound
engineer based in Washington State who decided to travel the
country in search of the few remaining quiet places. By his count,
the entire continental United States has fewer than a dozen sites
where you can't hear human-made noise for at least fifteen minutes at dawn. That's a pretty ridiculously low bar. But it is still so
out of reach. The quietest place in the country, Hempton discovered, is a spot in the Hoh Rainforest at Olympic National Park. If
you want to hear the earth without us, it's marked by a red stone on
a moss-covered log at 47-degrees 51.959N, 123-degrees 52.221W,

678 feet above sea level. But get there early; by midday, even there, you can hear overflights a dozen times per hour. Noise may well be the most pervasive pollutant in America.

I never thought much about airplane noise until I moved to D.C. I grew up on the eleventh floor of an apartment building in New York, where the sounds of the city were mostly muted and charismatic: a flash of mariachi, a distant ambulance, a summer storm. Out West, the planes were fewer and farther away. But my neighborhood now is one of the loudest in the city thanks to flights following the Potomac River as they roar in and out of Reagan National Airport. Jets fly overhead at a rate of about one every two minutes starting early in the morning, with average decibel levels between 55 and 60 but sometimes spiking much higher (60 decibels is high enough to drown out normal speech; over 80 can damage hearing).

I knew this moving in. Neighbors assured me I would learn to ignore the planes. "After a year or so, you don't hear them anymore," they'd said. But it's been over two years now and I still hear the planes. They drive me crazy. It's hard to eat alfresco, impossible to talk on the phone with the backdoor open. Between the planes and the routine security surveillance choppers, I feel like I'm in a militarized zone when I walk near the river. My gaze is drawn up, and I can read the logo on the fuselages. Sometimes, I can even make out the theme animal on the Frontier Airlines tail fins. There's the mustang! It's wildlife-viewing, D.C.-style.

Then there are the nettlesome sounds of competitive landscaping: the parading whines and drones of weed-whackers, lawn-mowers, leaf-blowers and, if I'm exceedingly unlucky and under deadline, circular saws. Such are the afflictions of close quarters, and they aren't necessarily new. The Victorian historian Thomas Carlyle didn't hear engines while working on his biography of Frederick the Great from his study in London, but he was made apoplectic by chickens, carriages and dogs. So maddened was he that he commissioned at great

88 FLORENCE WILLIAMSFLORENCE WILLIAMS

expense the making of a soundproof room in his attic. It nearly killed him. It was so airtight that when he lit up for a smoke, he passed out, only to be saved by the maid.

As Charles Montgomery writes in his book *Happy City*, "Living under the flight path of commuter jets is terrible for happiness . . . but we do not always respond logically to environmental stimulus." Right. The logical thing would be to go the hell back to Colorado. My neighbors aren't exactly wrong. People can become habituated to sound, at least partly. We've all heard stories of people who say they can't sleep if it's too quiet, or they can't work apart from a din. Some writers have apps that replicate the sounds of a coffee shop for when they are working at home. I know a New Yorker who now lives in the country, but he plays himself devotionally made recordings of 14th Street, sirens and all, to fall asleep at night.

I keep hoping this settling into noise will happen to me, that I will become inured or even nurtured somehow by the city sounds, but it isn't happening. In fact, I've learned that full habituation is a bit of a pipe dream. Just because you don't notice certain noises anymore doesn't mean your brain is not on some level responding to them. Scientists and regulators used to be interested in noise pollution because of the threat of hearing loss, which is real and happening to many of us at younger and younger ages. But even at dramatically lower volumes, noise poses risks far beyond our ear canals. In fascinating studies, people have been hooked up to electrocardiogram monitors while sleeping through plane, train and traffic noise. Whether or not they woke up, their sympathetic nervous systems reacted dramatically to the sounds, elevating their heart rates, blood pressure and respiration. In one study that lasted three weeks, the subjects showed no biological signs of habituating to the noise, and in another study that lasted for years, the biological effects only got worse.

———

THIS SUBCONSCIOUS VIGILANCE makes sense from an evolutionary standpoint. Sleeping or hibernating animals must still maintain
their capacity to react to danger. It's not uncommon in the animal
world for some species to lose their vision through evolution (like
those seriously ugly fish at the bottom of the ocean) or their sense
of smell (like dolphins, or, increasingly, humans), but there are no
known examples of evolution driving vertebrate species to lose hearing. This is our main "alerting" and "orienting" sense; it tells us not
only that something is out there but from which direction it's coming. Sound also triggers our strongest startle reactions.

Of course, nature didn't intend roaring jet aircraft to be processed by our nervous systems every sixty seconds. What does a loud
anthrophone do to us? The news is not good, not for us and not for
the birds, whales and other wildlife whose breeding and foraging
habits are upended by it. Numerous whale die-off events have been
attributed to navy sonar, the vibrations from which literally cause
heads to explode. In the remote backcountry of Yosemite National
Park, aircraft are audible 70 percent of the time, raising ambient
noise levels by about 5 decibels. That's enough to reduce the distance
at which prey species can hear a predator approaching by 45 percent. Lab experiments show that when female gray tree frogs hear
traffic noise, it takes them longer to find males who are calling to
mate, if they can find them at all. No backseat romance for them.

Sound is designed to be processed swiftly by the brain. Sound
waves travel through the air and collide with our eardrums, which
wiggle back and forth in response to volume and amplitude. Nerve
cells pick up these perturbations and send signals to our auditory
cortex, the brain stem and the cerebellum, which together process
fear, arousal and motion. As to the perennial question of whether
a tree falling in the forest makes a sound if no one is there to hear
it (first posed by Irish philosopher George Berkeley), the answer
is technically no. There is no sound apart from a sentient brain's

interpretation of molecules vibrating through air or water. The brain turns those molecules hitting the eardrums and pinnae into a mental idea of sound. Birds will hear the toppling tree, and fish will hear it too. But there is no thing called sound unless the vibrating molecules are processed into pitch.

Hearing evolved well before vocalization, and eventually became useful for communication. It's difficult to know which came first in evolution: the ability to hear or the ability to see, but fish are thought to have developed vibration-sensitive hairs hundreds of millions of years ago, before they could see. The fancy three-boned middle ear of mammals is—along with mammary glands—our defining trait. In the womb, we can hear before we can see. By birth, hearing is our most fully developed sense. Because sound waves vibrate through bones and the brain (the frequency of a violin note, for example, will cause neurons in the auditory cortex to fire at exactly that frequency) it is a sense we feel with our whole being.

It's only after sound signals wash through our limbic brains that the frontal cortex gets to weigh in, for example interpreting the big rumbles as a familiar DC-10, not a marauding lion. In the microseconds in between, though, a stress response has already begun. If, as Stanford neuroscientist Robert Sapolsky points out, lots of microstresses administered in a slow drip over time add up to chronic stress, then even something as harmless as airplanes heard during sleep can accrue in the stress bank.

Epidemiological and case-control studies overwhelmingly back up this observation. Many have been carried out in Europe, where high-density neighborhoods surround busy airports and where excellent health records are easy for researchers to access. In a study of 2,000 men over age 40, environmental noise above 50 decibels was associated with a 20 percent increase in hypertension. In another study of 4,800 adults over age 45, every 10-decibel increase in nighttime noise was linked to a 14 percent rise in hypertension. Health experts studying nearly a million people living near the

Bonn airport found that women living with noise over 46 decibels were twice as likely to be on medication for hypertension as those living with levels under 46 decibels. The World Health Organization attributes thousands of deaths per year in Europe to heart attack and stroke caused by high levels of background noise.

Researchers followed hundreds of children over two years before and after an international airport opened in Munich. They also looked at a control group of similar children who did not live as close to the airport. The stress hormones epinephrine and norepinephrine nearly doubled in the noisy-hood kids measured at six and eighteen months after the flights began. Their systolic blood pressure went up five points (the quieter-neighborhood kids' blood pressure went up two points).

In the largest and scariest study to date looking at noise pollution and children's cognition, funded by the European Union and published in the *Lancet* in 2005, researchers followed several thousand children attending elementary schools near major airports in the U.K., Spain and the Netherlands. They found significant impacts on reading comprehension, memory and hyperactivity. The results were linear: for every 5-decibel increase in noise, reading scores dropped the equivalent of a two-month delay, so that kids were almost a year behind in neighborhoods that were 20 decibels louder (results were adjusted for income and other factors). There's something real to the phrase "you can't hear yourself think."

As the authors of an important review paper on noise grimly noted: "The different types of stress reactions may... exert an adverse influence on the equilibrium of vital body functions. These include cardiovascular parameters such as blood pressure, cardiac function, serum cholesterol, triglycerides, and free fatty acids, hemostatic factors (fibrinogen) impeding the blood flow in terms of increased plasma viscosity... and presumably blood sugar concentration as well."

These health effects are serious. I'm frankly surprised they aren't better known, and that flight-path real-estate values don't seem to reflect them, at least not in D.C. After reading the studies, I loaded a decibel meter app on my phone. To my children's amusement, I've taken to running around and measuring the noise levels in and out of the house. Distressingly, they are comparable to levels associated with hypertension and learning delays in the studies I've been reading. I asked for noise-canceling headphones for Christmas, and I often wear them while working at home. Reagan National limits flights at night, but many international airports around the world don't. Technology offers some hope: jets have grown quieter in recent years and even muffled helicopters are being developed. Every decibel matters.

Interestingly, the researchers describe another outcome of hearing these noises: annoyance. It doesn't sound very scientific, but it turns out to play a big role in how people respond to noise, and therefore, stress. It's a simple concept: the more annoyed you are by the planes/trains/trucks, the worse you feel. Stress is not just a physiological response; it's a response that can be mediated by attitude or what psychologists sometimes call framing. This is why the adrenaline of skiing off a ledge into a steep chute can fill some people with energy, euphoria and focus and others with knee-buckling terror.

I realize this doesn't bode well for me regarding the airplanes, since I go out of my way to shake my fists at them. I just hope I don't become like eighty-two-year-old Frank Parduski, called "the world's first anti-noise martyr" by *New Scientist* magazine after he was run over by a motorcyclist he was harassing in order to get him to quiet down his two-stroke steed. But when visitors to national parks are told the loud airplanes overhead are part of important military exercises, many report being less disturbed by them. It's a good trick if you don't mind a dose of propaganda with your nature. It's not a plane; it's patriotism.

There's some evidence that more introverted or neurotic people are more annoyed by loud noises. They also may be less likely to become habituated to them. On the other hand, the louder and more intrusive the noise, the more likely you will grow annoyed. There's a bit of a chicken and egg problem. And whether you like planes or not, your brain still has to work hard to ignore them, and nobody can entirely Zen their way out of that.

THE U.S. NATIONAL PARK SERVICE is uncommonly interested in noise pollution because it operates under a federal mandate to protect its resources, including, since 2000, natural soundscapes. It's practically an impossible task, but as bioacoustical scientist Kurt Fristrup points out, a little bit of noise regulation can go a long way. Fristrup coordinates the science at the rather romantic-sounding "Natural Sounds and Night Skies" division of the agency. I imagine the staff running around wearing geeky headphones and glow-in-the-dark tee shirts depicting their favorite quasars. Fristrup's research agenda includes not only documenting the ill effects of anthropogenic noise on visitors and wildlife, but also documenting the beneficial effects of its absence: Why should we save the sounds of nature? What do they do for us? Fristrup is an accidental sound guy; he intended to study biomedical engineering at Harvard but got waylaid by paleontologist Stephen Jay Gould and evolutionary biologist E. O. Wilson. Biophilia rubbed off. Now he applies engineering to concepts of evolution, survival and ecosystem health. "We all interact with our environment through our senses," he told me, "so any pollution not only affects the fabric of our lives but our connections to everything else."

To learn more about how sound changes our brains and to find out just how noise-sensitive I am, I ventured to the sound labs of Pennsylvania State University. I was met by Peter Newman and Derrick Taff, two young park-rangers-turned-social-scientists in the Department of Recreation, Park and Tourism Management who work with Fristrup's group. Newman also didn't start out studying sound, he explained to me as we navigated a noisy cafeteria on campus. He was interested in parks and crowds, and was conducting visitor surveys at Muir Woods National Monument, known for its ancient redwoods.

"We asked if there was one thing to fix about the park unit, what would it be?" he explained. "And people said they wished it were more quiet. I was surprised what a big deal it was, but these were old-growth trees with a primeval feel, and visitors felt it should be quiet. Later we went back and analyzed the words they used, and they were so emotion-laden. Words like 'soothing,' 'peaceful.' That was interesting to us. That's where the research started dipping its toes into health." (And the survey carried weight: Muir Woods now has a "quiet zone," like the Amtrak quiet car: no phones, soft voices. It reduced the background noise there by three decibels, which is enough to double the listening area. So instead of hearing birds something like 10 yards in front of you, now you can hear them 20 yards away. That's a lot more birds.)

Now Newman and Taff run experiments out of the university's Acoustics Social Science Lab, the acronym of which, people noticed, resembles asshole, so they're switching the name around. Among other things, Newman and Taff and their colleagues have discovered that human-caused noise actually makes parks look worse, not just sound worse. Visitors hearing loud vehicle noise rate parks as 38 percent less scenic than those who don't hear it (and motorcycle sounds had the most impact, followed by snowmobiles and propeller planes). Counterintuitively, the soundscape was affecting the viewscape. Just imagine all the beauty we're missing out on. (Opposite effects are seen in cities, when people rate urban settings as more attractive when they can hear birdsong.)

Veering into human health, Newman and Taff decided to team up with Joshua Smyth, a biobehavioral health psychologist also at Penn State. He's interested less in how sound messes with your psyche and more interested in how it can make you feel better. Can some sounds be an intervention or an antidote for stress and depression? This appeals to Newman and Taff because natural sound is a resource the parks need to save before it's too late. If it's good for you, they want to know. They were familiar with the literature on nature as psychologically restorative, and it seemed to them that sound was a potentially powerful but underappreciated component of nature.

To tease out the sound piece, and to see how it worked for me, Smyth ran me through his current experiment. First he hooked me up to a heart rate monitor, which I would wear throughout. Then he gave me the Weinstein Noise Sensitivity Scale test, which asked a bunch of questions about my attitudes to various types of noise from things like a stereo to street traffic. I scored a 5.2. Adults average a 4, and college students average a 3.5, which puts me in the 88th percentile of sensitivity to noise. No surprise there. But in a short personality test, I emerged as not too neurotic, and of medium agreeableness (and no doubt more neurotic and less agreeable since moving to D.C.).

Next, I spit into a test tube to provide a reading of my pretest cortisol levels. Now the real fun would begin. In order to tell if nature sounds help "restore" subjects psychologically, Smyth has to first stress them out. Public speaking and math tests are two of the most dreaded tasks shared by a large number of people. So I was handed a pen and some paper and told to prepare a short speech about why I should be hired for my dream job. Partway through, my notes were abruptly taken away from me and I was told to stand and deliver the speech to a large mirror, behind which sat a panel of faceless judges. Several times during the 5-minute speech, I was interrupted and told to speak up. As I later discovered, this gauntlet of misery is called the Trier Social Stress Test (and it often includes a mental math component, typically repeatedly subtracting a number like 13 from a four-digit number). I figured Trier must be some sadist who devoted his life to freaking people out, but it turns out the test is named for Germany's University of Trier, where the test was formulated in 1993. It works: even though I knew there was no "panel of judges," I still showed a textbook response, with my heart rate climbing from the mid-60s to the mid-90s during the speech, and my cortisol levels (as revealed later) rising from 6.7 nanomoles per liter to 12.1. It's reductive to call cortisol a stress hormone, but lower levels generally mean lower stress. Researchers tussle over how reliable a measure this is (cortisol naturally varies over the course of the day, as well as during the menstrual cycle, so researchers often use it to study men).

Next, Smyth randomly assigns subjects to one of three recovery exercises: watching a fifteen-minute nature video with nature sounds, watching a fifteen-minute nature video with nature sounds and motorized sounds, or just sitting in a quiet room with no video. My video started playing, a simple scene from Yosemite of a summer meadow, some chirping birds, a blue sky. But a couple of minutes in, I heard a truck engine, followed by quiet, followed by the sound of a propeller plane. I'd been assigned to the second condition, and I again displayed a textbook response: once the nature video started, my heart rate immediately sank to baseline mid-60s range. When the truck rumbled, however, my heart rate shot up ten points. It took a while for it to drop again, but after more quiet nature, it plummeted down to the mid-50s. Now I was so relaxed I was practically dead. When noise #2 appeared, my heart rate shot back up, though not as high as the first time. My cortisol levels from this part of the experiment, at 8.2, reflected this almost-but-not-quite restored state (remember, my original level was 6.7 and my speech level was 12.1).

Smyth was also recording my heart-rate variability (HRV), which is fast becoming the darling of physiological stress measurements. It's increasingly used by scientists, medical doctors and athletic coaches. My HRV had also been monitored in Korea before and after hiking to tell me I had thickening veins. HRV is complicated to understand, especially in translation. It essentially measures—in real time—how quickly your autonomic nervous system responds to and recovers from microevents in the environment. Your heart is like a dancer—when it's relaxed, it swans up and down with fluidity. That's high variability, and it's good. But when you're stressed, that variability can clench into a much narrower range, the dancer getting a cramp. Some people have chronically low HRV, which is linked to a bunch of stress-related health outcomes like cardiovascular disease, metabolic disease and early death. During the speech test—and the loud noises—my HRV tightened up.

Noise, at least for me, really is a problem. The test showed that it's simply harder for someone who is noise-sensitive to fully unwind in an urban environment, regardless of its nice parks and nesting ducks. As Smyth put it: "Your recovery was clearly disrupted by the experience of noise. It set back your recovery with a carryover effect of at least a minute. For you, walking in the park, the benefits of nature may be offset by the noise of planes. Those noises are violating your experience of pleasant views and sound. It's half as stressful as doing the speech task. Those are aren't trivial effects."

Based on his research, Smyth has several recommendations for us sensitive types: try to reduce exposure to irksome noise through headphones, office insulation, etc.; if we can't do that, try to change our attitude about the noise—maybe by thinking that someday I will be on one of those planes getting the hell out of D.C.—and make an effort to experience positive sounds and quiet places.

"We should think about soundscapes as medicine," he said. "It's like a pill. You can prescribe sounds or a walk in the park in much the way we prescribe exercise. Do it twenty minutes a day as a lifetime approach, or you can do it as an acute stress intervention. When you're stressed, go to a quiet place."

In fact, Smyth thinks short nature-based interventions like this could help more people more efficiently than many other ones that get more attention, like meditation. "Meditation is getting all the glory. Unjustifiably," said Smyth. "Seventy percent of people will wash out." Not everyone likes nature, either, but just about everyone likes the noise to die down, at least occasionally.

THESE DAYS WE might worship absolute quiet, but John Ruskin wrote, "No air is sweet that is silent; it is only sweet when full of low currents of under sound—triplets of birds, and murmur and chirps of insects." To the extent that nature sounds are soothing to most humans, three in particular stand out: wind, water and birds. They are the trifecta of salubrious listening (favorite music and the voices of loved ones are perhaps the happiest of all, engaging almost every part of the brain, according to neuroscientist and musician Daniel Levitin, in *This is Your Brain on Music*).

Darwin devoted ten pages to birdsong and six to human music in *The Descent of Man*, noting that both have their origins in sexual selection, the desire to attract mates. As usual, he was correct. The Brits love birds so much that BBC radio broadcasts a daily ninety-second spot of birdsong. British Petroleum gas stations recently began playing birdsong in the bathrooms. "The aim was to create a mental connection with freshness," said a newspaper report. Good luck with that.

There appears to be something to the "freshness" idea. As British acoustics consultant Julian Treasure put it, birds sing in the morning, and we associate the sound with alertness and safety, a day when all is right with the world. This is how we've heard birdsong throughout our evolution. It's when you don't hear the birds that something is wrong. Also, birdsong is stochastic, random and non-repeating, so our brains interpret it not as a language but as a kind of background soundtrack. In fact, birdsong has some uncanny

similarities to human-made music, and its range and technical wiz-
ardry might, on some unconscious level, stimulate our happy-music
neurons. The French avant-garde composer Olivier Messiaen incor-
porated birdsong into his works and said of birds: "They are our
desire for light, for stars, for rainbows, and for jubilant song."

The brown thrasher can sing 2,000 songs. The cowbird has 40 different notes, and a horny chaffinch might sing half a million times
in a season. The Australian lyrebird is the world's best mimic, and
can imitate chainsaws, car alarms and the click of a camera shut-
ter (none of which reflects well on its habitat). The melodic hermit
thrush most often sings on a mathematic substrate that follows har-
monic intervals in recognizable pitches. The researcher who discovered this is named—I kid you not—Emily Doolittle, a composer at
Cornish College of the Arts in Seattle.

Despite the 300 million years that have passed since birds and
protomammals split from a common ancestor, our brains are surprisingly similar to the parts of birds' brains that hear, process
and make language. Humans share more genes governing speech
with songbirds than we do with other primates. This is because
humans and birds coevolved these language centers, both using
the same ancient neural hardware, specifically an area called the
arcopalladium in birds and the basal ganglia in humans, a region
also known for regulating emotion. It's well recognized that music
triggers emotions, but while much has been made of the ability of
Mozart to make us weep, tremble and rejoice (largely through the
release of dopamine in our mesolimbic reward pathway), birdsong
has received far less attention from neuroscientists.

Nevertheless, our doppelgänger birdbrain neurons may help
explain our primal affiliation to chirps, trills and tweets. In both
birds and humans, the ability to respond emotionally to linguistic
and musical sounds became mission critical for mating, communication and survival. The people who named Twitter knew what

they were doing. Psych studies using birdsong consistently show improvements in mood and mental alertness. An experiment at an elementary school in Liverpool found that students listening to birdsong were more attentive after lunch than students who didn't listen. Amsterdam's Schiphol Airport plays birdsong in a relaxation lounge that also features fake trees. People love it. Treasure, the British consultant, recommends that everyone listen to birdsong at least five minutes a day. I've been playing it on an app while writing this chapter. There's deep snow outside my window, but the spring birds are in full force on my phone. It does feel leavening. And my cat is certainly more awake.

"What I'm trying to do is figure out why it makes people feel better," said British environmental psychologist Eleanor Ratcliffe. Ratcliffe looked more like a high school student than a scientist. She had long red hair and wore a jean jacket that partially covered up a tattoo of parrots on her left arm. She admitted she was more of a city person than a nature person, but, as she put it, "one doesn't have to be in nature to be interested in it." I met her last summer for tea in the courtyard of the Victoria and Albert Museum, an excellent example of a restorative urban space. She opened her laptop, where tracks of birdsong were sandwiched between *The Sopranos* and a soul mix.

In her lab, she plays birdsong and asks subjects how they feel. "The overarching thing I'm finding is that people perceive bird sounds to be restorative, but it depends on the person, and it depends on the bird." Not all birds are loved equally. Many people dislike the raspy calls of jays and the brashness of crows and vultures. Ratcliffe launched into a disquisition the way an oenophile speaks of grapes. "Certain acoustic sounds, quiet, high pitch, bright and smooth are more restorative than loud and rough," she said. "The typical songbird, tweet tweet, the green finch or blackbird, robin, wren, have musical high trills. They are quite complex and melodious. It might help distract people from their troubles, but it's balanced between distraction and overwork. You want a bird that's not aggressive but submissive. Magpies are not restorative."

RATCLIFFE BELIEVES THAT sound can be restorative, and she's glad it's finally getting some attention in the research, but it's likely not the secret weapon of the nature cure. We're visual creatures, after all, and staring at a wall listening to headphones can take us only so far. Still, the lessons of sound can be translated in useful and creative ways. The city of Phoenix closes iconic South Park to vehicles one day a month for Silent Sunday. When I was in Korea, I'd gone for a walk along the Cheonggyecheon stream. "Stream" is a bit of an overreach. It's a stream in the way that Orange Julius comes from a tree or the Space Needle reaches space. The Cheonggyecheon used to be a ragtag underground ditch until it was unzipped to the world in 2005 as part of a greening initiative launched by Seoul's former mayor Lee Myung-Bak. To flesh it out, water is pumped in seven miles from another river and recirculated. Planted trees and flowering shrubs in the stream's canyon now attract insects and birds. The so-called "daylighting" of canals is one way for cities to make some nature visible again. In Seoul, though, one of its main purposes was to create a new soundscape to compete with the existing one of heavy traffic in the middle of the central business district.

At the entrance, a sleek waterfall drops down a generous story from street level, creating a pleasant rushing sound. At the bottom, I met Hong Jooyoung, a doctoral candidate in architectural acoustics from Hanyang University who specializes in using water sounds to obscure traffic noise. We walked along a good part of the three-mile-long watercourse, dodging other walkers, joggers and picnickers. Some young women were standing around looking at pigeons on the bank. It was a good place to hang out. Among its many benefits, the path here is six degrees cooler than the roadway above in the height of summer. Only about 20 feet wide, the stream often flows over rocks and through reeds. It literally burbles and whooshes, its soothing sounds amplified by the stone walls lining the sunken ribbon of water and path. Hong explained to me that with these new water features, it's the perception of traffic noise that changes. You can still hear the noise, but you don't notice it anymore. The traffic here is loud, above 65 decibels, but so is the water. "The creek design maximizes the sound," he said. "People don't think of it as noisy because it's a nice noise. They rate this kind of water sound as most favorable."

I was reminded of something the National Park Service's Kurt Fristrup had said, that unless we learn to make cities sound better, we stand at risk of losing the range of this precious sense. He calls our tendency to wear earbuds during all hours of the day "learned deafness." We are tuning out the real world in favor of our own personal soundscapes. The cost is we forget how to listen. And we lose an opportunity for true mental restoration.

"It's this gift we are born with, to reach out and hear all these incredible subtle sounds," he'd said, "and it's in danger of being lost in a generational amnesia. Some ears will never get a chance to develop sensitivity to those sounds."

Although Seoul's creek plan initially drew opposition because of its cost—about $380 million—and the need to reroute an elevated highway, it is now exceedingly popular, visited by thousands every day. The mayor went on to become South Korea's president.

ON THE LAST morning of my short vacation in Maine, I woke up very early and snuck out of my stepmother's house while the kids were still sleeping. I donned the EEG cap and slid into a kayak and onto a small lake. One on side sat a rural subdivision, boats and docks; on the other bloomed a generous expanse of the White Mountain National Forest. I paddled through a foot of soft mist resting on the water's surface. I couldn't see my blade as it touched the water, but I could hear the drips, and the birds on the approaching far bank. Occasional jets flew overhead, but they seemed very far away. A car started up down at the far end of the lake. Not too bad. It was quite peaceful. I filled my lungs with the mist and the sun and the bird-song, and I regally paddled onward in my proud EEG crown.

The morning's software algorithm report read like a Trekkie horoscope: "In most people, the alpha rhythm is attenuated when the brain is busy processing and responding to visual stimuli. How-ever, your brain produced substantial alpha even with your eyes open, suggesting that your brain dynamics are governed by long-range cortical connections and that you enter a relaxed state very easily."

Hah! I got alpha! I'd finally tricked the machine into thinking I was some sort of yogi. For a few moments on a quiet lake, I was.

Box of Rain

[When] the myopia had become stationary, change of air—a sea
voyage if possible—should be prescribed.

—HENRY EDWARD JULER, *A HANDBOOK OF OPHTHALMIC
SCIENCE AND PRACTICE*, 1904

She promised us south rooms with a view close together, instead
of which here are north rooms, looking into a courtyard, and a
long way apart. Oh, Lucy!

—E. M. FORSTER, *A ROOM WITH A VIEW*

One of the serious risks of city living are other drivers. Although our brains have long been hardwired to fear snakes and spiders, they are remarkably less attuned to the dangers of two-ton vehicles. Instead of dreaming about things that slither in the night, we really should be having nightmares about Yellow Cab, but the Freudians wouldn't have nearly as much fun. Two years ago, my seventy-five-year-old father was walking to work in downtown Silver Spring, Maryland, when he was struck by a car traveling 35 miles per hour. The accident was probably a combination of inattentive walking and inattentive driving, although my father was found solely at fault because he wasn't in the crosswalk.

In the intensive care unit at Bethesda Suburban, the nurses were shaking their heads. This was the third pedestrian accident they'd seen that week. In D.C. alone, there are over 800 such accidents a year and the number is rising despite more speed cameras. Dad suffered seven broken bones and a traumatic brain injury, and nobody could predict how well, or if, he'd recover. At first, he looked good, still tan and strong in the starchy, space-age hospital unit as though he'd mistakenly landed on the wrong stage set, but that soon changed. He was in terrible pain, unable to eat, and very confused. He couldn't understand language and he was capable of muttering only the phrase "condo fee" over and over. He didn't know where he was and he kept trying to pull his various tubes out and bolt. He was, in the unexpected lingo of the hospital, "an elopement risk."

I'd already lost one parent and I didn't want to lose another. After two weeks in the ICU, he was transferred to a rehabilitation hospital known for achievements in neurology. Because of its high concentration of medical researchers, facilities and experience with everyone from returning veterans to gunshot victims, Washington, D.C., is an excellent place to have a brain injury. The belief is that if you rehab early and hard, you can recover much function.

This is the man who taught me to love nature, to cross rivers by jumping on rocks, to lean my weight out while scampering down a boulder, to tack a sunfish and to steady a canoe. This is the man who, even in New York City, would scurry us up to the bleak, tar-covered roof to watch the orange sun dip beyond the Hudson River. Every year for Christmas, he made me a book about our wilderness trips the previous summer. They were filled with grainy images of river rapids and rock cliffs. The one from 1978 is titled "Adventurous." In his acknowledgments, he calls me out. "This is specially written for her. It is printed in a limited edition with only one copy." For a long time these books were sort of painful in an embarrassing way for me to read. My father's earnestness, his sentimentality, my eye-rolling adolescence. But reading them now, I find they are full of insight into our divorced family and the role that the natural world played in his mental landscape.

In 1979, I was twelve and Dad was in the midst of a difficult relationship with a girlfriend. We spent a couple of weeks paddling the wilderness lakes around the Canada-Minnesota border. A picture from that trip shows us sitting on a broad boulder by the shore, sharing a huge loaf of bread. I am wearing my new Swiss Army knife on a lanyard around my waist. My father, deep in his Grape Nuts phase, is tan and lithe, bearded, long-haired and shirtless. "This year more than ever finding extraordinary solace in these odysseys with my daughter," he wrote that year. "Early in the trip, my head was still full of dilemmas to be resolved. I was less accessible, more quick to anger. Yet as the events of the trip developed, my anxieties became less severe and I started to feel some measure of balance. I felt a peace such as I had not known for many months. What is it about me and water?"

Dad grew up climbing trees in Richmond, Virginia, and tending the family's victory garden. Blessed with good health his whole life, he was never long without walks or other adventures in nature. Now this had changed. There are few places farther removed from natural landscapes than a typical hospital room. Because I was researching this book at the time of his accident, I knew enough to request a bed near a window for his long stay in rehab.

I had, for example, come across Florence Nightingale's famous nursing textbook from 155 years ago: "It is the unqualified result of all my experience with the sick, that second only to their need of fresh air is their need of light," she wrote. "It is a curious thing to observe how almost all patients lie with their faces turned to the light, exactly as plants always make their way towards the light." I'd read Oliver Sacks's account of recovering from a serious leg injury after he'd fallen down a cliff in Norway while being chased by a bull (not all writers live such exciting lives). After many weeks in the hospital he finally went outside, where he would "fondle the living plants. Some essential connection and communion with nature was re-established after the horrible isolation and alienation I had known. Some part of me came alive." Even if my father couldn't name the objects he could see, the sunlight and the trees and the birdsong might somehow reach him.

We've looked at smell and sound. Now it's time to tackle our strongest sensory system for processing the world around us: the visual. Its impact on our emotional and physiological states can also be immediate and powerful. One of the first people to study the health consequences of a room with a view was psychologist and architect Roger Ulrich, the researcher who wondered in the mid-1980s why people went out of their way to drive on tree-lined roads and who measured alpha brain waves in subjects looking at nature slides. After those initial, promising results, he was curious about effects in the real world, so he turned to a suburban hospital in Pennsylvania. Like Sacks, he knew from personal experience that nature could play a role in healing. As a child, he suffered recurring bouts of painful kidney disease. During long periods at home in bed, he drew great, inexplicable comfort from a pine tree outside his window. Later, as a young scientist, he wanted to test his hypothesis that nature views could reduce patient stress and lead to better clinical outcomes. He was aware of a study from 1981 showing that prisoners in Michigan whose cells faced rolling farmland and trees (instead of a barren courtyard on the other side of the facility) had fewer sick-call visits overall.

Ulrich examined the records of gallbladder-surgery patients over half a dozen years, some of whom had been assigned to rooms with a window view of trees and some who looked out onto a brick wall. He found that the patients with the green views needed fewer postoperative days in the hospital, requested less pain medication and were described in nurses' notes as having better attitudes. Published in *Science* in 1984, the study made a splash and has been cited by thousands of researchers. If you've ever noticed a nature photograph on the ceiling or walls of your dentist's exam room, you have Ulrich to thank.

SINCE THEN, WINDOW STUDIES have examined everything from schools to office buildings to housing projects. They have shown that nature views support increased worker productivity, less job stress, higher academic grades and test scores and less aggression in inner-city residents. The studies measure something different and far less ambitious than a full sensory immersion in a hinoki forest. They look at "accidental nature," the exposure you get without trying. It's the mere blot of green glimpsed on the way to the laundry or between sentence diagrams. Some of the studies are small and seem vulnerable to confounding factors. Perhaps people who are wealthier, healthier and happier to begin with prefer to be closer to nature? The best studies, though, are large and designed to weed out competing factors.

Frances Kuo, yet another academic spawn of the Kaplans at Michigan, is a psychologist who now runs the Landscape and Human Health Laboratory at the University of Illinois, Urbana-Champaign. She was interested in constructing experiments to test the logical playing-out of Kaplans' Attention Restoration Theory. If our brains get fatigued by too much direct attention, and if that makes us irritable, then wouldn't we also be more likely to become violent? Could spending time looking at nature make us less violent, and if so, would a simple view out a window be enough to make a difference? Among her seminal studies were some from the early 2000s looking at views, violence and cognition at the brutalist Robert Taylor housing project (now razed) in Chicago. Some of the buildings faced barren asphalt streetscapes and some faced modest lawns dotted with trees. Residents were randomly assigned to apartments and shared equally dismal levels of poverty, drug use, education attainment and employment status. It was a perfect window-view laboratory.

Kuo and her colleague, William Sullivan, interviewed 145 female residents (most of the units were occupied by single mothers) and found that those with the asphalt views reported higher levels of psychological aggression, mild violence and severe violence than their tree-view counterparts. In a separate study, the asphalt viewers also reported more procrastination behaviors and assessed their life challenges as more severe and longer lasting. Kuo and Sullivan knew that aggression is linked to impulsivity, so they undertook another study of children in the Robert Taylor complex. They found that those living with the barren views were less able to control impulsive behavior, resist distractions and delay gratification. The results applied to girls but not to boys, which Kuo attributed to the fact that the girls were likely spending more time indoors where the views mattered. Because these findings were based on questionnaires, Kuo and Sullivan wanted a more objective measure, so next they turned to police reports. These were tied to a different Chicago housing project, Ida B. Wells, which was distinguished by a series of courtyards ranging from no greenery to mixed concrete/greenery to a pretty lush landscape with grass and trees. Analyzing 98 buildings over two years, they found a striking correlation between the level of greenery and the number of assaults, homicides, vehicle thefts, burglary and arson. Compared to buildings with low amounts of vegetation, those with medium levels experienced 42 percent fewer total crimes, and the contrast between lowest and highest levels of vegetation was even more pronounced. Buildings with the most green views saw 48 percent fewer property crimes and 56 percent fewer violent crimes than buildings with the least greenery.

Kuo didn't think it was the greenery alone that was magically lulling people into peace and harmony; rather, in the case of Ida B. Wells, it was that the prettier courtyards drew residents outside, where they got to know each other and could keep an eye out. The researchers had also tested how often residents used the courtyards and asked them what they thought of their neighbors. The greener-courtyard residents reported their neighbors were more concerned with helping and supporting one another, had stronger feelings of belonging, engaged in more social activities and had more visitors.

The Kuo findings were backed up by a Dutch study of over 10,000 households that found people of similar incomes living near more vegetation experienced less loneliness, and by an office study showing that subjects in rooms with potted plants were more generous to others when asked to distribute five dollars than those in a room without plants. (Potted plants! Someone really ought to deck out the halls of Congress with ficuses.) For some reason, social psychologists like to study road rage, and even here, the evidence for tree views making us nicer appears strong. In these studies and in others, the greenery appeared to be leading to prosocial behaviors and a stronger sense of community. Frederick Law Olmsted suspected as much.

"I am not historically a nature lover," Kuo told me. "I had no personal intuition when I started that these findings would come out the way they have. But twenty years later, I have convinced myself."

ALTHOUGH THESE STUDIES point to real health and behavioral effects from nearby nature, they don't explain how merely looking at some shrubbery—as opposed to a full sensory immersion in nature—makes us healthier and nicer. For that, the visuals need to be broken down. Enter nanoparticle physicist Richard Taylor. Like Ulrich's, his quest starts with a meaningful childhood experience. When Taylor was ten years old and growing up in England, he chanced upon a catalogue of Jackson Pollock paintings. He was mesmerized, or perhaps a better word is Pollockized. Franz Mesmer, the crackpot eighteenth-century physician, posited the existence of animal magnetism between inanimate and animate objects.

Pollock's abstractions also seemed to elicit a certain mental state in the viewer. Today, in his fifties, Taylor is positively da Vincian in his range of pursuits—besides his day job in nanoparticle physics, he is also a painter and photographer with two art degrees—but his long, curly hair looks more Newtonian. His hair is so remarkable that the University of Oregon, where he works, once Photoshopped it out of a publication. Perhaps the marketing department considered it a distraction, as Eugene isn't exactly known for conservative dress standards. Come to think of it, my high school physics teacher had exactly this hairstyle. Must be a thing.

Taylor never lost his interest—obsession, really—in Jackson Pollock. While at the Manchester School of Art, he built a rickety pendulum that splattered paint when the wind blew because he wanted to see how "nature" painted and if it ended up looking like a Pollock (it did). He made his way to Oregon's physics department to study the most efficient ways to move electricity: in multiple tributaries like those found in river systems, or lung bronchi, or cortical neurons. When electrical currents move through things like televisions, the march of electrons is orderly. But in newer tiny devices that might be only a hundred times larger than an atom, the order of currents breaks down. It is more like ordered chaos. The patterns of the currents, like those branches in lungs and neurons, are actually fractal, which means they repeat at different scales. Now he's using "bioinspiration" to design a better solar panel. If nature's solar panels—trees and plants—are branched, why not manufactured panels? He frequently paddles around Eugene's Waldo Lake when he's chewing on a problem.

Several years ago Taylor wrote an essay describing a seminal insight: "The more I looked at fractal patterns, the more I was reminded of Pollock's poured paintings. And when I looked at his paintings, I noticed that the paint splatters seemed to spread across his canvases like the flow of electricity through our devices." Using instruments designed to measure electrical currents, he examined a series of Pollocks and found that the paintings were indeed fractal. It was a little like discovering your favorite aunt speaks a secret, ancient language. "Pollock painted nature's fractals twenty-five years ahead of their scientific discovery!" He published the finding in the journal *Nature* in 1999, creating a stir in the worlds of both art and physics.

Benoit Mandelbrot first coined the term "fractal" in 1975, discovering that simple mathematic rules apply to a vast array of things that looked visually complex or chaotic. As he proved, fractal patterns were often found in nature's roughness—in clouds, coastlines, plant leaves, ocean waves, the rise and fall of the Nile River, and in the clustering of galaxies. To understand fractal patterns at different scales, picture a trunk of a tree and a branch: they might contain the same angles as that same branch and a smaller branch, as well as the converging veins of the leaf on that branch. And so on. You can have fractals within chaos, or you can have fractals creating what looks like chaos. When I look at the equations describing these relationships, my eyeballs spin, but to a mathematician they are clear, consistent and beautiful. Arthur C. Clarke described the Mandelbrot set (a beetlelike drawing that illustrates these equations) as being "one of the most astonishing discoveries in the entire history of mathematics."

Although true fractal patterns occur quite commonly in landscapes, in space and in living creatures, even potato mold, they are rare in abstract art. So rare that when a trove of previously unknown paintings was discovered in a storage locker belonging to a family friend of Pollock's in 2002, Taylor was called in to verify their authenticity. There was much at stake. If the paintings were really Pollocks, they were worth hundreds of millions of dollars. Taylor's computer analysis showed the paintings did not in fact exhibit Pollock's signature fractal geometry. The physicist concluded they were fake. It was a bold and controversial assessment, but later validated when chemical analysis proved some of the paints were manufactured too recently to be used by the artist, much to Taylor's relief. Fractals had interrupted one of the boldest forgery plots of all time.

Taylor was curious to know if there was a scientific reason people love Pollocks so much. Was it the same reason everyone was installing fractals as screen savers and flocking to stoner light shows at the planetarium? Could great works of art really be reduced to some eye-pleasing nonlinear equation? Only a physicist would dare ask. If this breed is not daunted by the origins of the universe, it certainly isn't by abstract expressionism. So Taylor ran experiments to gauge people's physiological response to viewing images with similar fractal geometries. The early work was funded by NASA, which wanted to decorate space stations with stress-reducing images (but, interestingly, not images that reminded astronauts of faraway Earth, because that would be too sad-making). Taylor measured people's skin conductance and found that they recovered from stress 60 percent better when viewing computer images with a mathematical fractal dimension (called D) of between 1.3 and 1.5. D measures the ratio of the large, coarse patterns (the coastline seen from a plane, the main trunk of a tree, Pollock's big-sweep splatters) to the fine ones (dunes, rocks, branches, leaves, Pollock's micro flick splatters). Fractal dimension is typically notated as a number between 1 and 2; the more complex the image, the higher the D.

After the NASA work, Taylor went deeper. He and Caroline Hagerhäll, a Swedish environmental psychologist with a specialty in human aesthetic perception, converted a series of nature photos into a simplistic representation of land forms' fractal silhouettes against the sky. They found that people overwhelmingly preferred images with a low to mid-range D (between 1.3 and 1.5). Did preference reflect some sort of mental state? To find out, they used EEG to measure people's brain waves while viewing geometric fractal images. They discovered that in that same dimensional "magic zone," the subjects' frontal lobes easily produced those elusive and prized alpha brain waves of a wakefully relaxed state. This occurred even when people looked at the images for only one minute. EEG measures waves, or electrical frequency, but it doesn't precisely map the active real estate in the brain. For that, Taylor has now turned to functional MRI, which shows exactly the parts of the brain working hardest by following the blood flow. Preliminary results show that mid-range fractals activate some brain regions that you might expect, such as the ventrolateral cortex (involved with high-level visual processing) and the dorsolateral cortex, which codes spatial long-term memory. But these fractals also engage the parahippocampus, which is involved with regulating emotions and is also highly active while listening to music. To Taylor, this is a cool finding. "We were delighted to find [mid-range fractals] are similar to music," he said. In other words, looking at an ocean might have a similar effect on us emotionally as listening to Brahms.

To hear Taylor describe it, Pollock was actually painting nature in his abstractions, the natural law of fractals. Taylor believes our brains recognize that kinship to the natural world, and they do it fast. Pollock's favored dimension is similar to trees, snowflakes and mineral veins. "We've analyzed the Pollock patterns with computers and compared them to forests, and they are exactly the same," said Taylor. This dimension does more than lull us; it can engage us, awe us and make us self-reflect. "Furthermore," explained Taylor, "the exposure only has to be 'environmental'—they don't need to stare directly at the pattern. A person will receive the effect, for example, walking down a corridor with the patterns on the wall." Or, presumably, working by a window. Taylor does not know how long these positive effects last, but he's working with medical researchers to see whether it's possible to restore some brain functionality in stroke victims by exposing them to fractals.

But why is the mid-range of D (remember, that's the ratio of large to small patterns) so magical and so highly preferred among most people? What, for example, leads people like my father to warble in one of his homemade books: "Big raindrops hit the water making symmetrical patterns of crosses surrounded by bubbles. Surreal and very moving. The quiet visual effects are making the patterns of the world seem very different. It is as if to experience the world in a new way . . . not with words but with images."

Many patterns in nature fall into the low-to-mid range, including clouds and landscapes. Taylor and Hagerhäll have an interesting theory, and it doesn't necessarily have to do with a romantic yearning for Arcadia. In addition to lungs, capillaries and neurons, another human system is branched into fractals: the movement of the eye's retina. When Taylor and Hagerhäll used an eye-tracking machine to measure precisely where people's pupils were focusing on projected images (of Pollock paintings, for example, but also other things), he saw that the pupils used a search pattern that was itself fractal. The eyes first scanned the big elements in the scene and then made micro passes in smaller versions of the big scans, and it does this in a mid-range D. Interestingly, if you draw a line over the tracks animals make to forage food such as albatrosses surveying the ocean, you also get this fractal pattern of search trajectories. It's simply an efficient search strategy, said Taylor. Other scientists have found this D range elicits our best, fastest ability to name and perceive objects, something our brains do when facing new visual information. This is a critical task; we need to assess quickly what's friendly and what's dangerous, among other things. If a scene is too complicated, like a city intersection, we can't easily take it all in, and that in turn leads to some discomfort, even if subconsciously. It makes sense that our visual cortex would feel most at home among the most common natural features we evolved alongside, like raindrops falling on a lake.

"Your visual system is in some way hardwired to understand fractals," said Taylor. "The stress-reduction is triggered by a physiological resonance that occurs when the fractal structure of the eye matches that of the fractal image being viewed." So perhaps our comfort in nature is not really about an innate love for living things or the physical frisson of a good view— it's simply about fluent visual processing. It's about an easy congruence in the way the outside stimulus (the tree) is processed internally by our neurons. Taylor uses the word "resonance" instead of congruence, which is interesting, because it's the same word Beethoven used to describe how he felt when he left the confines of Vienna for the country, which I also quoted in the introduction: "How happy I am to be able to walk among the shrubs, the trees, the woods, the grass and the rocks! For the woods, the trees and the rocks give man the resonance he needs." Long before fractals, Beethoven intuited a powerful alignment of senses and surroundings.

According to this processing theory, if the cause of our relaxation is not rooted in Arcadian romance, the solution surely is. We need these natural patterns to look at, and we're not getting enough of them, said Taylor. As we increasingly surround ourselves with straight Euclidean built environments, we risk losing our connection to the natural stress-reducer that is visual fluency. For a lot of reasons, it would be good to bring greenery back to cities and get outside. But Taylor has already begun to think about solutions beyond parks or looking out the window. "You don't always have a window with a view. We may be able to manipulate and fool the visual system and come up with an even better range [of fractal dimension] than nature, purify it and maximize the response," he said, beginning to sound a little scary. As if sensing my response, he added, "I don't want some Orwellian future where you project a perfect fractal in a public space and everyone must stare at it for five minutes. But we want to give this information to architects and artists so they can integrate it into a variety of works."

In sensing the existence of an energy force between objects and people, perhaps Mesmer wasn't such a crackpot after all. I had one final question for Taylor. I was interviewing him via Skype video because he was on holiday in Australia. His soft curls tumbled to the lower edges of the screen like a fine galloping creek.

"Is your hair fractal?"

He roared with laughter. "I suspect my hair is fractal. The big question of course is whether it induces positive physiological changes in the observer!" I believe it may have.

MY FATHER DID recover, slowly and then quickly, amazingly, in his sun-filled semiprivate room with a view. He saw physical therapists, speech therapists, occupational therapists, lots of family who chattered to him and urged him to talk back. There was clearly more than nature at work on his battered brain. Of course my elbowing him into a bed near a window meant his roommate wasn't near the window. There aren't enough windows to go around, and even when there are, sometimes the views don't cooperate. Perhaps Taylor had a point. Wouldn't it be handy if you could just turn on a video screen of a glade or fractal waterfall, or even just slap a poster on the wall?

That is one conceit, anyway, being explored at the maximum-security unit of the Snake River Correctional Institution in eastern Oregon. In a unique experiment in partnership with social scientists, the prison staff has agreed to play nature videos in the exercise room of one wing of the prison. The cells in Snake River offer no windows at all, and the only "outdoor" courtyard is tiny and surrounded by buildings. Its only view is the sky through a grate. Snake River is a difficult place: it has a higher-than-usual percentage of inmates who commit suicide and self-harm, and it's not unusual for staff to perform "cell extractions" on those who are out of control, kicking and screaming and banging on doors. Prisoners in solitary confinement are perhaps the most nature-deprived people on the planet. They are often mentally ill when they enter prison, and become more so as the weeks and months tick by.

But now inmates can lift weights and do chin-ups several times a week in the so-called Blue Room while watching forty-minute videos of ocean life, rainforests and desert sunsets. Since the Blue Room went in two years ago, inmates often request to go in there when they want to calm down. Said Renee Smith, Snake River's behavioral health services manager, "We're getting plenty of stories from officers saying they feel like it is relieving stress and mental health and behavioral issues. We're feeling that they're not getting into trouble as much. We feel like there are less cell extractions, less hollering and screaming."

But how close is virtual nature to the real thing? Wondering if the screens could in fact have the same stress-lowering effects, a psychologist named Peter Kahn at the University of Washington ran a couple of experiments at his university. In the first, he placed nature-playing video screens in windowless offices and found that they did improve workers' cognition and mood. In the second, he divided ninety subjects into three groups: one with a real-live window view of nature, one in front of a plasma-screen TV showing nature, and one near a blank wall. He first distressed the volunteers with public-speech tasks and then measured how quickly each group recovered. Taken together, the studies showed that the real-nature views helped the most, with the video views helping a bit (although hardly at all in the second experiment) and the blank wall helping the least. Kahn concluded that humans can "adapt to the loss of actual nature," but "we will suffer physical and psychological costs."

While some researchers like Kahn lament this speedy and inexorable replacement of real nature by screens, others, especially the younger ones, seem more pragmatic. They also, notably, grew up with less exposure to nature to begin with. "We are moving toward more of a virtual life with every year, with video games, 3D TVs, larger, more immersive screens and more virtual content," said Deltcho Valtchanov, a twenty-something postdoc in cognitive neuroscience at the University of Waterloo in Ontario who grew up in the urban core playing video games. Valtchanov came to the topic not because he was interested in nature or art, but because he was interested in its antipode, technology. He wanted to validate, or even ennoble, virtual reality, to prove that it could elicit "real" nervous system activity. His university review board wouldn't let him instill fear in human subjects, so he started reading the dusty psych literature on what made people feel relaxed instead, and he landed upon nature. This was a surprise to him, and he didn't really believe it, not being much of a nature guy himself. But it worked so well to soothe subjects in his master's degree experiments that for his Ph.D. research, he decided to try to deconstruct the visuals to figure out why. The ultimate goal would be to make the virtual-reality experience even better. Because if you could, there is no end to what a couple of nerdy guys with a headset can do. "Why wouldn't you escape your real life?" asked Valtchanov. "This way, you can enjoy your own living room and it's relatively cheap. You can go to Hawaii without the bugs and the jet lag."

WHEN I LEARNED that Valtchanov had eventually developed a smartphone app that could rate and categorize nature scenes and then, ultimately, synthesize them, I had to check it out. He had recently completed his doctoral work here on the featureless plain of southern Ontario. When I visited on a gray, windy February day, I could see how it might inspire VR. It also evidently inspires tech of all flavors. Although most Americans have never heard of it, many Silicon Valley gurus consider Waterloo to be their best feeder school, topping even Stanford. Valtchanov, dressed in black jeans, a checked button-down shirt and sporting a soul patch, led me through windowless serpentine hallways in the basement of the psych building. We passed a small room with photorealistic bright blue, cloud-speckled ceiling panels, manufactured by a company called Sky Factory, whose motto is "Illusions of Nature." "Wouldn't it be nice to have this in your house instead of lights?" he asked. "Wake up and turn the sky on?"

I guess, I figured, but then again, I like to actually look out a window. But there was no time to debate; we moved on to the Research Laboratory for Immersive Virtual Environments, optimistically if not ironically dubbed ReLIVE. The room is cinderblock with concrete floors, about 14 by 20 feet.

Here, he would introduce me to his scientifically derived restorative world. He wired me up to finger electrodes for measuring my galvanic skin response (GSR, otherwise known as sweat) and an infrared sensor for my heart rate. He asked me to calculate out in my head the answer to 13 times 17, and then 12 times 14. On cue, I immediately stressed out. Then he crowned me with a precision-tracker 3D headset, a bit like scuba goggles but tricked out with a gyroscope and accelerometer. This would capture my movement so the 3D video could respond, fully immersing my brain in Valtchanov's virtual paradise. At least that's the idea.

A generously sized Samsung monitor fired up, and I found myself walking, or rather, walk-floating, on a deserted island in the tropics. Valtchanov creates these worlds over thousands of hours, adding sounds like birds, water streaming, chirps, grass rustling, the thud when we jump off small rises. The movement was strange. Valtchanov was controlling my speed and direction, so I felt like I was being dragged by my forehead through an environment at high speed.

"Do you feel like you're the game-master guy in *The Hunger Games*?" I asked him, half expecting balls of flame to start smacking me.

Valtchanov virtual-pulled me along a path, my virtual feet crunching on the ground, then down a hill, through some tall grasses, then to a beach. I started getting woozy. Then I was suddenly dragged underwater for a few moments, which I don't think was supposed to happen.

I couldn't help but feel a little alarmed. Were there sharks? Were there spiky urchins to step on? Is bad weather rolling in? It didn't really feel relaxing to me. I told Valtchanov.

"Not all nature is restorative," he said. "Being in tall grasses is not necessarily a nice thing. But can you hear the ocean? We're going to head toward a waterfall, and there's a rainbow there."

But I was not going to enjoy Valtchanov's rainbow.

I felt like I was about to throw up.

Later, after I took a break to hyperventilate in the bathroom and splash cold water on my face, Valtchanov told me what I already knew. I didn't do well at virtual relaxing.

"YOUR GSR DID not go down," he said, disappointed. "It stayed where it was. Maybe that was the motion sickness. I apologize. The technology is getting better for that, so you don't feel like you're watching through someone else's eyeballs." I wasn't alone, he explained. He had to throw out 30 percent of his data because of subjects approaching the puke zone. This has been a major hang-up in the development and marketing of consumer VR. "The motion sickness is due to the technology being old," he said. "It's being solved by better displays that don't have that ghosting. When you turn your head quickly, you'll notice edges blur."

Yes I did. Bummer. But I was also secretly a little proud. I was one of those remaining holdouts for whom only the authentic experience will do. My skepticism for the virtual approach carried over to Valtchanov's app, called EnviroPulse, which was still in beta testing. A bit like a magic kettle, you put an image in, such as a window view, and watch a number come out predicting your emotions. Can't we predict our own responses to a particular view?

Obviously not, responded Valtchanov, although politely. If so, why would we build such ugly cities and suburbs, schools and hospitals? It's not the views we mischaracterize, it's our responses to them. We walk right past magnificence all the time, not just because we're busy, or because we don't see it, but because we don't realize what it's capable of doing to our brains. Valtchanov is here to help. He envisions a Yelp-like, crowd-powered app that can make recommendations for the most relaxing outcrop in Central Park or the best route to take to work. "Instead of looking for food you can look for happiness," he said.

Here's how it works: You hold your phone up to a scene, or a photograph, and the app puts it through a series of algorithms to judge its restorative potential. Natural images contain statistics. Fractals, as Valtchanov explained, are just one of them. Color is important, as is saturation, shapes (humans prefer rounded contours to straight lines), the complexity of the contours, and luminescence (we rate brighter, more saturated colors as more pleasurable). All of these visual properties have been studied over the years for their emotional weight, and these data feed the algorithms. For example, it's well known that the colors red and orange excite or agitate people (and make us lustful and hungry, as purveyors of fast food well know), while blues, greens and purples tend to relax us. The human eye is well designed to respond immediately to color. In our retinas, we have three color-sensing types of cone cells primed to pick up reds, blues and greens, and those cones enjoy a direct line to the brain's visual cortex, a spot of geography in the back of the head. Most mammals possess only two types of cones (and can't distinguish between red and green), but primates, being the visual monopolists we are, are special in this regard (we have three cones). But not overly special. Some creatures, like birds and butterflies, have five cones, enabling them to see technicolor infrareds and ultraviolets. The mantis shrimp trumps us all, sporting somewhere between twelve and sixteen cones. God knows what they see, but it must be trippy.

Colors help us spot and distinguish foods and notice things out of the ordinary. Red pops out at us because we have more cone cells dedicated to picking up this color, and in many cultures, red was the earliest color given a name after black and white. Since red makes us vigilant and energized, we walk faster down red corridors than blue ones. As the English philosopher Nicholas Humphrey has said, "If you want to make a point, say it in red." When Olympic boxers and martial artists wear red, they win more often. But pink, interestingly, has the opposite effect, weakening athletes, making prisoners less aggressive (hence the color known as drunk tank pink) and pacifying psychiatric patients. In a study where agitated hospital patients looked at a blue light, their tremors subsided.

Based on the literature on sensory perception, Valtchanov's app gives blue the highest score of all. Predators tend not to be green or blue. Biophilia proponents would argue we've learned to associate these colors with life-giving, healthy ecosystems full of plants (green), clean water (blue) and expansive reflection (sky azures, ocean teals). Since we all live under that sky and drink its offerings, these hues may instill feelings of universality and shared humanity. Similarly, as John Berger writes in *The Sense of Sight*, "That we find a crystal or a poppy beautiful means that we are less alone, that we are more deeply inserted into existence than the course of a single life would lead us to believe."

I'm drawn to the rich intersections of culture and science to be found in color, but it's spatial frequency that gets Valtchanov most excited. He's convinced it's this—regardless of the fractal content— that unlocks the doors to paradise. Spatial frequency captures the complexity of contours, shadows and shapes in a scene or image. We prefer images that are easier and faster to understand.

In the app, straight and jagged lines are rated very low on the restoration scale compared to smooth and rounded ones. "Urban jagged edges are not so good for you," said Valtchanov. But like Taylor, he believes there's a Goldilocks sweet spot of complexity, not too busy and not too boring. For his Ph.D., Valtchanov used an eye tracking machine to parse how people looked at scenes. He found that while the eyes tend to linger lazily over nature scenes, urban scenes provoke many more rapid "fixations," and more blinking, indicating that the eyes—and brain—are working harder to decode them. These places demand our attention.

From his research, Valtchanov believes easy-to-process scenes trigger the release of natural opiates in the brain. Other studies have shown that images we love activate a primitive part of the brain called the ventral striatum (strongly linked to deep emotions and rewards that motivate our behavior) as well as the opioid-rich parahippocampus—the same region Taylor found stimulated in subjects viewing fractals. When the poet and writer Diane Ackerman writes of craving the "visual opium" of a sunset, she is not being as metaphorical as she thinks. According to Valtchanov, nature makes us happy because of a neural mechanism in our ventral visual pathway that is tuned to a mid-level frequency range like a clear radio signal. When it finds it, happy molecules flow.

This is the brain spot Valtchanov wants to target with his app. To show me how it works, we pulled up a bunch of images on the Internet. We held up the phone to the photographs and watched as a small bar on the image moved like a thermometer from green (good) to white (neutral) to red (stressful). The app will also give the image an absolute score of restorativeness between 0 and 100 and code them to these colors. Some of the ratings were predictable. Forest vale: very green. Lake: ditto. Urban intersections: red. Simple buildings: neutral. Shanghai skyline under blue sky: neutral. But when I pulled up a snowy meadow flanked by a snow-covered peak, the kind you would see on a travel brochure for the Rockies, the app went to reddish.

"What's up with that?" I asked.

"Well, it's jagged and it's white and the trees look dead, because it's winter."

"But it's beautiful," I said. "When I'm skiing in places like this, I'm definitely in my happy place."

"The app isn't taking into account your activity or endorphins or oxygen to your brain. I'm just analyzing the face value of the environment. According to Wilson's biophilia hypothesis, people would react strongly to dead trees."

"But these aren't dead. It's just winter. It's pretty."

"There's a difference between pretty and psychologically valuable." He adjusted my hands in front of the image. "If you point the camera a bit upwards to get more of the blue sky, it will rate better." He shrugged. "I'm not saying it's perfect."

TAYLOR, VALTCHANOV AND OTHERS have shown that nature images—even on a screen—can elicit fast, positive responses in our brains. But if nature, real nature, is what the visual system was actually built to look at, maybe we should let those looks linger. Because when we're stuck indoors looking at screens, our eyes aren't happy. Mine get dry and start to hurt. I went to my eye doctor for eye pain, and she was like, welcome to the club. "You're a starer." She told me. "A starer?" I suddenly felt like a creepy ogler. "You don't blink!" she said. I blinked. I blinked again. It felt weird. "When we stare at screens all day, we blink less," she said. "We all do it." She sent me off with some eye drops and told me to make myself blink twenty times in a row as often as I can remember.

Aside from dryness, weird things start happening to our eyes in the absence of outdoor space and light. One clue was a study from China that found twice the rates of myopia (nearsightedness) in wealthier, urban parts of the country than in rural areas. In Shanghai, a stupendous 86 percent of high school students need eyeglasses. As recent studies in Ohio, Singapore and Australia found, the real difference between those with myopia and those without is the number of hours they spend outside. Sunlight stimulates the release of dopamine from the retina, which in turn appears to prevent the eyeball from growing too oblong. Indoor and outdoor light are totally different beasts. Even on overcast days, outdoor light is ten times brighter and covers vastly more of the light spectrum. Educators are scrambling to come up with solutions, including installing full-spectrum indoor lights and glass ceilings over classrooms.

There's a better solution: go outside.

I find the intellectual compulsion to break apart the pieces of nature and examine them one by one both interesting and troubling. I understand it's the way science typically works: to understand a system, you have to understand the parts, find the mechanism, put your flag on a piece of new ground. The poets would find this is nonsense. It's not just the smell of a cypress, or the sound of the birds, or the color green that unlocks the pathway to health in our brains. We're full sensory beings, or at least we were once built to be. Isn't it possible that it's only when you open all the doors—literally and figuratively—that the real magic happens?

For that, you need more than a few moments on a screen or in nature. You need, to be exact, five hours a month.

PART THREE

FIVE HOURS A MONTH

You May Squat Down
and Feel a Plant

The faint whisper of rain and running water was still there and it
had the same tender note of solitude and perfection.

—TOVE JANSSON

Once upon a time in Finland, there were little forest spirits who could put spells on people who were too noisy or who treated the forest with disrespect. The victims would experience a condition called *metsänpeitto*, which translates as being "covered by the forest." In this state they suddenly found themselves unable to get their bearings. Nothing looked familiar. A kind of intense fascination would overcome them. They could hallucinate and experience supernatural phenomena.

Long after the birth of Christ, strong pagan beliefs continued in the boreal lands between the Baltic and North seas. *Metsänpeitto* is well documented into the nineteenth century, and, like other religious experiences, was more commonly experienced by women and children. The celebrated Finnish poet V.A. Koskenniemi dedicated a poem to the condition in 1930. It is a favorite of Marko Leppänen, a journalist and activist, who read it aloud to me in sonorous, incomprehensible Finnish on a small island in the Helsinki archipelago.

"*Metsänpeitto* is not necessarily negative," explained Leppänen, a tall, lean, smooth-skinned man in green woolens standing over a stunted pine. "*Metsänpeitto* is about getting lost in beauty. It could have a taste of freedom, nature-union and joy. The poem is suggesting that."

In other words, *metsänpeitto* is a little like forest-bathing on acid. It's very Finnish. It's also the opposite of the short-term window-view effect of nature; it represents a deeper surrender to the forces of the forest. Many health experts here believe modern times call for a full, if still only occasional, immersion in nature. They're trying to figure out how much time outdoors is needed for healthy, ordinary citizens to stay sane.

Leppänen is fascinated by the mind-altering, health-giving effects of wildish landscapes, and he wants to share them with others who visit him on the island of Vartiosaari. One of many small cones of forested bedrock emerging from the Baltic Sea, the rugged isle lies within Helsinki's city limits. In winter, people walk across the sea ice to get here (and nearly every year someone falls through and drowns). By the time I arrived on a sunny day in May, the ice had melted and we took a quick dinghy ride.

Leppänen, who appears ageless but is actually forty-four, is the island's unofficial groundskeeper, druid and spokesperson. Amid the ferns, pines and craggy sea cliffs on the tiny island sit a dozen or so houses, a grid of garden plots, and, thanks to Leppänen, a nature trail. Considered a rogue nature preserve, Vartiosaari hosts an unusually rich collection of woody plant species in a variety of landscapes. "The whole island is only eighty-three hectares, yet it feels much larger," said Leppänen. Many people manage to get lost here, but they seem to be happy after many hours of being lost. I think it's a health effect to get lost."

In the early twentieth century, a managing director of Nokia (then a wood pulp and rubber company) liked the island of Vartiosaari so much that he quit his job to live there, building a house called Quisisana, from the Latin, meaning "where one heals." To enhance the island's salutary attributes and create more momentum to protect the place from encroaching development, Leppänen cobbled together some funding from the Finnish Forest Research Institute and the city of Helsinki and marked out a "health nature trail," complete with signposts, recommended exercises and descriptions.

This isn't your typical park fitness trail. Our first stop was a big gray boulder, a glacial erratic that toppled off an iceberg when the island was once underwater. The far-traveling rock, said Leppänen, reminds us of the importance of moving, of exercise. It's a metaphysical StairMaster. We walked on a few paces and arrived at a small outdoor chapel featuring a stone altar, a timbered cross and bark-sided benches to remind us of spirituality in nature. Next we considered a mutant pine tree, growing outward at waist-height instead of growing upward. Leppänen called it "the table of Tapio" after a Finnish forest god. "This can be for our offerings, a symbol of gratitude," he said. "To be grateful is good for your health. Today we can be grateful to ourselves for visiting this forest!" We walked along to a stone-laid labyrinth the size of a large living room. This was constructed by locals in 1999, but it's a nod to an ancient islander tradition. No one's really sure what the old labyrinths were for, but to Leppänen they represent mystery, wandering and play.

This is about the time it struck me that the Finland of grown-ups is not unlike my daughter's old Waldorf preschool in Boulder, complete with paganistic rites, woodcrafts and Middle-earth symbology (in fact, J. R. R. Tolkein was reportedly influenced by the Kalevala, a Finnish creation epic in which the world is born from the cracked egg of a diving duck). The group I was hiking with even broke for a snack circle. They didn't start singing or making headpieces out of twigs, but I could see it coming.

To the Finnish, being outdoors in nature isn't about paying homage to nature or to ourselves, the way it tends to be for Americans. We fetishize our life lists, catalog peaks bagged and capture the pristine scenes of grand wilderness. It is largely an individual experience. For the Finnish, though, nature is about expressing a close-knit collective identity. Nature is where they can exult in their nationalistic obsessions of berry-picking, mushrooming, fishing, lake swimming and Nordic skiing. They don't watch moose; they eat them the way their ancestors did. And they do these things often.

According to large surveys, the average Finn engages in nature-based recreation two to three times per week. Fifty-eight percent of Finns go berry-picking, 35 percent cross-country ski, often in Arctic darkness, under lights in large city parks. Seventy percent hike regularly, compared to the European and American average of about 30 percent. Fifty percent of Finns ride bikes, 20 percent jog and 30 percent walk a dog, and I particularly like this one: 5 percent of the population, or 250,000 people, partake in long-distance ice-skating. All told, over 95 percent of Finns regularly spend time recreating in the outdoors.

It could be that the Finnish exist in something of an arrested state of development, or perhaps the rest of us somehow got overdeveloped. We put down our floral wreaths earlier, acting, for better or worse, like civilized grown-ups. Finland is highly unique among Western countries for urbanizing very late in the game.

"It wasn't until the 1960s and '70s that masses of people finally went to cities. Before that we were forest people," said Leppänen as we walked the soft forest paths. "We haven't had opportunity to escape nature. It's very thin, this urban layer. You can still today see, we are walking here in the capital city and it's seven kilometers to the heart of the city, yet this could be from hundreds of kilometers away. This is an intact nature landscape. It could be different, if we were living many generations in an urban setting." To him, civilization is like the spring sea ice, transparent, the wild pulse below still sensate.

Being just two generations removed from the land—and being a nation with few immigrants—means that nearly everyone still has a grandparent on a farm or woodlot. Those grandparents still live in country houses, or they own a modest, seasonal country house even if they've moved to the city. Finland has 5 million people, and 2 million *kesämökki*, or "summer cottages," so almost every family still has a rural, nature-based anchor. It's a middle-class real estate paradise.

Finland scores high on global scales of happiness. Many people assume this is because there isn't much income disparity here. But perhaps it's also because everyone has access to what makes them happy—a bunch of lakes, forests and coastlines, combined with ridiculously long, state-sanctioned vacations and a midnight sun. (Of course, there is a flipside, the grim, dark winters, when Finns drink too much and act up, unless they're skiing.)

Like many Finnish Gen-Xers, Marko Leppänen grew up chasing butterflies. He spent nights in trees by himself as an eleven-year-old while his American counterparts were playing Pac-Man in suburban split-levels where the only moss was the color of the shag carpet.

Until recently, Finns have lived off the land, both emotionally and economically. Sure, Finland came up with the flip phone, Angry Birds and the wildly popular set of comics by Tove Jansson built around Moomin the talking snowman. But the nation's dominant industry is forest products, in the form of renewable fuel for clean-burning energy plants and paper pulp. Finland is the most forested country in Europe, with trees covering 74 percent of the land. As one visiting British journalist noted, "the view was a bit samey." The forests are mostly privately owned in small holdings, but, mirabile dictu, at least to an American mindset, there is virtually no such thing as trespassing. Finnish law operates under the concept of *jokamiehenoikeus*, or "everyman's right," which means anyone can traipse over anyone else's land, picking berries, picking mushrooms, picking their nose, whatever. They can even camp and make campfires. They only things they can't do are cut timber or hunt game. (Right-to-roam laws in a few other aggressively democratic European countries such as Denmark, Norway and Scotland are similar but not quite as lenient.)

To many Americans, this sounds like a socialist takeover of private property (contrast these laws to the "my castle" laws in states like Montana, where you actually have the protected right to shoot trespassers dead). To the Finnish, though, *jokamiehenoikeus* is the essence of freedom, because it means you can walk forever. In a small country where everyone is distantly related, the please-share-nicely concept works.

It makes sense, then, that the Finnish are uniquely devoted to their forests, and are coughing up cash to study them if for no other reason than to justify their constitutionally protected frolicking. Although they do have other motivations, and some of them we can relate to: the Finns report increasing levels of stress, depression and obesity as they move into urban environments. That national recreation survey that mentioned long-distance ice-skating also noted that, in almost all categories, frequency of outdoor activities has dropped in the last ten years, no doubt replaced by staring at brightly lit devices inside their houses. Even the Finns can't resist them.

The country has some choices to make. If time in forests can be shown to reduce health-care costs, improve mental health and promote fitness, planners can use that information to argue against paving places like Vartiosaari as Helsinki grows. Even if we think the Finns are gnomish outliers, we can likely learn a few things from what researchers here have discovered.

———

LIISA TYRVÄINEN FREQUENTS a Helsinki restaurant called Kaarna, which means "Bark," as in tree, not dog. She used to be an ecologist, but she got tired of feeling that her research didn't really matter to planners and policy-makers, so she got a Ph.D. in economics. She studied how things like forest and park views dramatically increase housing values. "The phenomenon of what nature means to Finnish politicians is all about how to valuate it," she said while giving me a tour of Helsinki's parks. She became intrigued by the research out of Japan indicating that forests had concrete physiological effects on human health. In a country like Finland, which is trying to figure out how to manage its vast forests for the benefit of people and industry, the health piece, if real, seemed like it could be another useful column in the national spreadsheets. Is it worth saving natural areas or not? "I'm wanting more data. I don't want to be part of rubbish research, hugging trees," she said.

Now Tyrväinen runs a research division at the National Resources Institute of Finland, a government-funded agency. She visited Japan and then invited some of the *shinrin yoku* researchers over to Finland to advise her on setting up similar experiments. She had some issues with the Japanese protocols and wanted to tweak the experimental design. Miyazaki and his colleagues were mostly studying young Japanese men in small groups. Tryväinen wanted bigger studies and better controls. In the Japanese experiment I observed, for instance, one group was loaded into a van and driven a couple of hours to a park, while the other group went straight to downtown. It's possible that some of the lower blood pressure and cortisol levels attributed to "nature" could just be the result of more time to space out on the drive.

Tyrväinen secured close to $16 million for a series of studies known as the Green Health and Research Project. In Tyrväinen's Japan-inspired studies, all participants sat in a van for the same amount of time and they included more women, more adults, and more office workers. Also, the Japan team studied hard-core urban vs. hard-core nature. Tyrväinen wanted to look at environments available to everyone in the city: a busy street, a managed city park, and a more wild forest park. The managed park resembles parts of New York's Central Park that are manicured and landscaped, such as the boat pond and surrounding meadows. The forest park, Helsinki's beloved Central Park, reminds me of the deep parts of the Ramble but with bigger, taller pines and some straight avenues.

Tyrväinen also wanted to measure blood pressure because of its known links to stress and disease. "It's the long-term physiological benefits we're interested in. We'd like to follow these people." And she was hunting for more granular information: "What is an optimal amount, location, type and size of nature spaces for health in everyday living environments?"

Tyrväinen's team is interested in what ails normal working people and what helps them. Their aim is not to improve productivity per se but to lower national health-care costs and to provide city planners with data for managing green space. If she can help make people feel better, that's fine too, but she's an economist, not a social worker. In Europe, 60 percent of job-related health problems are, like bad backs, musculoskeletal. But the next-highest category (14 percent) is psychological: stress, depression and anxiety. The Finnish call it "burnout syndrome," and it significantly taxes both employers and government health agencies.

I had to guffaw a bit when I heard about Finnish worker stress. The Finns typically work eight-hour days. About 80 percent of workers are unionized. They get five-week vacations, pensions and health care, as well as one-year paid parental leave (men as well as women are encouraged to take time off). When I was sending scores of emails overseas for this book, I would frequently receive messages that the recipient was on parental leave for the next several seasons and not checking email. If these workers are stressed out, what did this bode for Americans, 25 percent of whom get no paid vacation at all?

The Finnish government is funding Tyrväinen because it knows it has a limited pool of workers in a small country. As her colleague Jessica de Bloom told me, "In other countries, you select the right person for the job and if that person gets burned out, then you find another person. Here, you keep that individual as long as possible, you keep them happy."

So while the Japanese researchers had given their subjects questionnaires about mood, Tyrväinen's team decided to add other quantifiable measures of restoration, vitality and creativity, all related to happiness on the job. If the Kaplans' Attention Restoration Theory is correct, the Finns would expect to see higher scores after time in nature. Sample questions for restoration (participants are supposed to rate the statements on a scale): "I feel calm." "I have enthusiasm and energy for everyday routines." "I feel focused and alert." Sample question for vitality: "I feel alive and vital." And for creativity: "I got several new ideas." While self-answered questionnaires aren't as sexy or reliable as objective measures of brain waves and hormone levels (sometimes the participants can guess what the researchers are after, potentially biasing results), in larger studies they tend to be pretty accurate, especially in conjunction with other types of physiological or cognitive tests.

In one study, Tyrväinen and her colleagues asked 3,000 city dwellers about their emotional and restorative experiences in nature. They found the biggest boosts occurred after five hours a month in natural settings. Tyrväinen wanted to drill deeper into the data, so for another study, her team took 82 office workers, mostly women, to each of the three different sites: city center, manicured park and forest park. At each place, before and after sitting for fifteen minutes and then after walking leisurely for thirty minutes, the researchers collected questionnaires, saliva samples, blood-pressure and heart-rate data. Throughout, the volunteers were instructed not to speak to each other (to eliminate the positive psychological benefits of socializing). If people felt happy, it would not be from making friends.

The results turned out to be what scientists call beautiful. There were significant effects and linear dose responses that followed predictions. Compared to sitting in the van, the volunteers did not feel psychologically "restored" in the city, but they did in the park and forest. They experienced these changes relatively quickly, after just fifteen minutes of sitting outside. After the short walk, these restorative feelings continued to increase. The more time people spent in the green areas, the better they reported feeling, and the effects were slightly stronger for those in the wilder forest. But the benefits weren't just about relaxation; on measures of vitality, which you'd think might rise in the city, only nature did the trick, although it took the full forty-five minutes. Both the vitality and restoration scores dropped in the city, to the point where participants in the park or forest felt 20 percent better than their urban peers. The greenies also felt stronger positive emotions and lower negative emotions, and the respondents reported feeling more creative. On the objective measures, cortisol levels dropped in all three settings, perhaps a result of being away from work demands, speculated Tyrväinen.

The good news for city dwellers is that just fifteen to forty-five minutes in a city park, even one with pavement, crowds and some street noise, were enough to improve mood, vitality and feelings of restoration.

"The results of our experiment suggest that the large urban parks (more than 5 hectares) and large urban woodlands have positive well-being effects on urban inhabitants, and in particular for healthy middle-aged women," the study concluded, as published in the *Journal of Environmental Psychology*. The results supported the earlier five-hours-a-month recommendation. But the research-ers also noted the dose-response relationship: the more nature, the better you feel. To elevate mood and stave off depression most reli-ably, Tyrväinen told me, "five hours per month is the lowest amount of time to get the effect, then after, if you can go for ten hours, you will reach a new level of feeling better and better."

I did some quick calculations. Five hours per month means get-ting out there in the verdure a couple of times a week for about thirty minutes. To achieve ten hours a month requires spending about thirty minutes in nature five days per week. Or, as one of Tryväin-en's colleagues told me, "two to three days per month outside the city would bring the same effect." No wonder country houses are so popular; the Finnish nervous system needs them. The Finnish-approved nature cure won't work for everyone, because these results reflect averages. But in a country with a high proportion of mildly depressed people, if it works for even a small percentage it will translate into huge savings for the national health-care system.

And in Finland, parks and woodlands are an easy solution. "Nature here is cheap and free for everyone," said Tryväinen.

IF TYRVÄINEN IS interested in valuing forests for the sake of the Finnish economy, one of her collaborators, Kalevi Korpela, is moti-vated by a desire to boost the dark Nordic psyche. The Finnish word for healthy, *terve,* derives from the word for "hardy pine," able to withstand storms. Finns withstand a lot: long, dark winters, freez-ing temperatures, a collective historic memory of being regularly invaded and colonized by Swedes and Russians. From the Swedes, they learned brooding. From the Russians, they learned drinking. The Finns themselves are notoriously taciturn, introverted and a bit shy. One study found that of many nationalities in the world, the Finns are the most comfortable with long silences. They are not chatty. There's been much discussion of the Scandinavian paradox:

countries like Sweden, Denmark and Finland rank very high on happiness indexes, but they also suffer high rates of suicide.

Korpela's grandfather fought the brutal winter battles of World War II and, like so many survivors of his generation, ended up suffering silently. Nobody knew how to talk to these broken men about their pain, which was immortalized in classics like Väinö Linna's *The Unknown Soldier*, Finland's all-time best-selling novel. Korpela, an experimental psychologist at the University of Tampere, has spent most of the last twenty years studying how different environments make us feel. Unusual for psychologists two decades ago, he was most drawn to positive psychology, or what made us feel good. From his experiences during childhood, when he and his older brother had the run of the town while their parents worked long hours, he knew that place mattered to his own psyche and might for others as well.

Tampere itself is not terribly impressive geographically. A city of about 250,000 people ninety minutes north of Helsinki by train, it was founded by Swedish King Gustav III at the relatively late date of 1779. The city sits along a set of rapids—now corralled into a hydro dam—on the Tammerkoski River. Overlooking the city is the highest esker in the world. (I didn't know what an esker is either—it's basically a glacial moraine.) This feature is more like a geological speed bump than a mountain, rising only 85 meters. The fact that the Finns are so proud of it tells you what you need to know about the country's topography. You won't find majestic peaks or canyons. Marshes are so predominant here that the country gets 9 percent of its electricity from peat gas. Finland is the Saudi Arabia of peat. Still, the close connection between people and the land is evident from Korpela's own life and work.

"As a teenager, I used to sometimes go run in the woods and stop at a big rock where you could see the lake," he said. "I noticed it was a way of calming myself and regulating my emotions so I had this habit of going and stopping there." Now a trim professor sporting facial hair reminiscent of Freud's goatee period, Korpela has become known for studies about "favorite places" and their positive influence on mental health. In his studies, when he asks respondents to name their favorite places, over 60 percent describe a natural area such as lake, beach, park, garden or woods.

If there was something special about nature, Korpela wanted to find out how quickly it worked on our emotional brains. If the psychoevolutionary theory of Roger Ulrich (hospital window guy) is correct, then our responses to pleasant nature spots should be automatic, and perhaps immediate. One classic way to measure positive and negative emotions is to show people pictures of faces and have them rate them for moods like fear, anger, happiness and surprise, while timing the exercise. Happier people will recognize happiness in others more quickly, and take longer to recognize fear or disgust.

Korpela primed a group of volunteers by quickly showing them photographs of various scenes that had been manipulated along a spectrum from urban to buildings-with-trees to just trees or parkland without structures. After each photo, the volunteers were asked to identify the emotions in pictures of the faces he showed them. Interestingly, after looking at scenes with more nature, the subjects were quicker to recognize happiness and slower to recognize negative emotions like anger and fear. The inverse was true after the more urban shots. In other words, looking at nature photos made them behave (instantly) in happier ways. To Korpela, the study confirmed Ulrich's hypothesis of nature causing a rapid emotional response at a subconscious level.

As we've seen in Part One, nature appears to have some immediate effects: a lower pulse rate and the beginnings of a parasympathetic nervous system response leading to feelings of peace and well-being. Korpela scoured the literature and came up with a sort of time-response matrix. Thanks to his faces study, he knew the

quickest responses: "Within 200 milliseconds, people react positively when they see images of nature," he explained. "The picture
you've seen affects how you respond, because the picture evokes
your emotions." Moving up the matrix, Ulrich's experiments with
the bloody woodshop videos followed by nature videos showed a
decrease in subjects' heart rates, in facial muscle tension, and
changes in skin conductance typically occurring within 4 to 7 minutes. The Japanese and Finnish studies found lower blood pressure,
lower circulating cortisol and improved mood after 15 to 20 minutes. At around 45 or 50 minutes of being in nature, many subjects
show stronger cognitive performance as well as feelings of vitality
and psychological reflection. What if Korpela could thread all these
observations together in a way that enhanced the effects in a real-
world application?

He came up with the idea of a "Power Trail," a well-signed, self-
guided nature walk that maximizes nature's beneficial effects.
Hikers wouldn't need a specially certified ranger or a class or a big
healing forest, just some views, ideally including water, and strategic instructions. In 2010, the Ikaalinen Spa in central Finland
let Korpela construct a trail network around its property with
government funding (and about the word "spa": lest you think it
connotes an exclusive enclave for ladies in lululemon, you should
know that, in Finland, spa visits are a federal benefit for workers in need. Yet another reason to brave the sea ice and move to
Finland).

The trail was an immediate success, according to Korpela and
Tyrväinen, and now there are half a dozen similar ones throughout northern Europe. They surveyed the hikers who use them, and
found that 79 percent said their moods had improved, with greater
boosts in those who walked the longer loop (6.6 kilometers) than the
shorter loop (4.4 km). Gender, age and, interestingly, weather had
no effect on the results. But they also found that about 15 to 20 per-

cent of people just don't dig it. These people may hate bugs, or the sky, or whatever, and no matter how biophilic their brains are supposed to be, they simply can't relax in nature.

To test it out for myself, I headed out to spa-ville with Korpela in his silver Peugeot. To be honest, it was sort of relaxing from the get-go. I was also experiencing what social scientists call the novelty effect, in which things that are new and fresh can make us feel good. This is why we like to travel, peruse the photos in *National Geographic* and even fall in love serially. I was in love with the lack of midweek traffic in rural Finland. It was May, and so we passed rolling fields of canola flowers, young corn and wheat. We stopped for lunch at a café in a log house that was painted baby blue. We grazed from a buffet featuring slabs of moose with lingonberries. The novelty effect was in full swing.

Once settled into the spa's parking lot, Korpela pulled out a blood-pressure machine. I sat silently for two minutes and then measured my levels, which were already in the mellow zone. Leaving Korpela to his own personal Power Trail moment, I set out on the path, which meandered past the spa's wood-burning saunas, around a lake, and literally over hill and dale. It was a walk in the country, pleasant but not spectacular. There were birds and blossoms and trees along with a few houses and tractors and woodpiles. Being alone, said Korpela, is a good way to maximize certain benefits, especially the ones having to do with self-reflection. Of course, the Finns would say being alone in nature is best; they are notoriously introverted. But thirty years ago, the psychologist Joachim Wohlwill agreed, writing that natural environments experienced in solitude seemed especially restorative to people who are mentally fatigued or socially stressed. I get it. I love being alone in nature when it feels safe. (Women, not surprisingly, tend to rate being alone in nature as more stressful than men do, because of concerns for safety.)

Right after setting out, I came to a sign on the trail, marking the first of nine stations. I pulled out a piece of paper with Korpela's English translations. Station one was a cognitive task: it showed two line drawings of a busy picnic scene around a lake. I was to find and count all the differences between the two images. For example, one included a woodpecker on a tree limb while the other showed no woodpecker. There was also a brief questionnaire asking me to rank how I felt on a scale of 1 to 5. This is called the Restorative Outcome Scale, frequently carted out for psych studies. The statements include "I am feeling calm and relaxed," "I am alert and focused," "I'm enthusiastic and energetic" and "All my everyday worries are away." I'd repeat both tasks at the end of the hike and compare my scores.

Farther along, station two sported a sign instructing me to look at the ground and the sky, breathe deeply and relax my shoulders. "Feel your mind and body becoming calm," it said. When I looked up, I saw power lines, which deflated me, until I remembered that this trail is lit for winter skiing. That made me happy again.

Station three asked walkers to listen to the sounds of nature and "let your thoughts run free." Also, "you may squat down and feel a plant." Station four asked me to walk to a spot nearby where I feel peaceful. Station five: identify your mood and state of mind. And so on, through to finding an element of nature from the view in front of you that could be a metaphor for yourself. I chose a tall tree sheltering smaller trees. I missed my kids and was getting sentimental now.

At the end of the walk, I retook the cognitive test and questionnaire. If you score more than ten points higher on the scale, the interpretive sign essentially tells you that you need to get your butt into nature as often as you possibly can. If your scores were the same or lower, you should just go home and turn on some European football. I scored five points higher, which meant "this kind of walking suits you and you should try it again sometime," translated Korpela.

The whole exercise felt a bit like taking a personality quiz in the back of *Mademoiselle*. "What Does your Favorite Snack Food Say About You?" Or from the Internet: "Which Muppet Are You?" Psychological questionnaires gained popularity in the 1920s, when Carl Jung was writing about personality types. Not sure Jung had Kermit in mind, but people love these tests. If they get people out hiking more, so much the better.

My cognitive test scores and my blood-pressure results were more inconclusive. My compare-the-illustrations scores were the same. My systolic pressure dropped quite a bit—six points—but my diastolic went up nine points. A lot of things affect blood pressure, including states of hydration, so I'd call it a question mark. My heart rate, though, dropped a point. I was relaxed before the hike and still relaxed after it. For now, I was off to sip some calendula tea and sample Finnish chocolate from a farm café. I was beginning to wonder if reporting about the pleasures of nature was making me too mentally stable to be a reliable research subject.

But for stressed-out workers, Korpela sees quick, regular visits to green space as having enormous potential to relieve the daily grind. Based on his studies, he said "a thirty-to-forty-minute walk seems to be enough for physiological changes and mood changes and probably for attention."

The five-hours-a-month recommendation stands for those of us in need of a short tonic and as a way to ward off everyday blahs. But what if you're not just a frazzled worker? What if you've got bigger problems? It would be up to the Scots and the Swedes to figure out how to get already seriously depressed people into the woods and gardens and make them stay there for a while. Twelve weeks ought to do it.

Garden of Hedon

Clearings. That's what I needed. Slowly my brain righted itself
into spaces unused for months.

—HELEN MACDONALD

n the Gaelic poem "Hallaig," by Sorley MacLean, a man is forced to
leave his favorite grove of trees for America during the land clear-
ances of the nineteenth century. This poem, worshiped by so many
in Scotland, speaks directly to the national soul in its tragedy, sen-
timent and land-love. "I'm finding it difficult not to cry when I think
about it, and I'm English," an ecologist named Peter Higgins told
me. The landscape here, as in Finland, is a unifying force, rooted in
the bones of people who grew up with it. It's also rooted in the Gaelic
language itself. There's the word weet, to rain slightly, and williwaw,
a sudden, violent squall, and wewire, to flit about as foliage does in
wind, and that's just the *W*'s. How perfect is this: crizzle, "the sound
and action of open water as it freezes"?

For all that landed pride, though, Scotland is a country divided in
ways that places like Finland and South Korea are not. It is divided
not just over the perennial question of whether to cleave from
England. The urban poor are unmoored from the land, and from

Scotland's deep culture of resilience. Some would argue the two are related. Consequently, the country's attitude toward nature has a more desperate tinge; the survival of a culture and of a people are in play. The idea of spending more time outdoors is emerging as an important tool for regaining health and sanity already lost.

Nowhere is the country's social divide more evident than in Glasgow. Upon arriving, I was immediately struck by the down-and-out vibe just below my hotel. Edinburgh is all lovingly pre-served stone architecture, uni students rushing about, tourists buying tweed, and Harry Potter fans taking selfies in front of the Elephant House café, where J. K. Rowling did some scribbling. But downtown Glasgow recalls the Bowery of the 1930s: sleeveless drunks in the middle of the day, young people smoking sullenly on the streets, Here, the underclass is largely white, hopped up, and pissed off.

Parts of Greater Glasgow face the lowest life expectancies in all of the European Union. In some neighborhoods a man can expect to live to 54, while 12 miles away he will live to 82. Sixty percent of the city's excess deaths are triggered by just four things—drugs, alcohol, suicide and violence. Alcohol-related deaths increased four-fold between 1991 and 2002. The main cause: economic disparities driven by four generations of unemployment following the disman-tling of manufacturing and mining in the 1970s and 1980s.

It's this divide that gets Richard Mitchell, an English epidemi-ologist at the University of Glasgow, up in the morning. While the Finnish and Japanese nature studies targeted the educated middle class, Mitchell is looking at the beaten-down poor. He's spent years researching effective messaging for preventing alcoholism and obe-sity. Now, though, he's turned to the environment itself. Long fas-cinated by why some places breed healthy people and some places don't, he was intrigued by research in the Netherlands to start look-ing at maps of green space. Dutch studies had shown remarkable mental and physical health benefits of living within half a mile of green space, including reductions in diabetes, chronic pain and even migraines. Mitchell wondered if one of the main reasons for the association was simply exercise.

This assumption makes sense. When we are out in nature, we are generally self-propelled, breathing in oxygen, liberating our lungs and our cardiac capillaries from their usual cramped, desk-hunched configurations, and arresting, temporarily, the slow backward death march of our telomeres. Exercise as a cure for all things has been so drilled into the public health establishment that it crowds out everything else, with the possible exception of quitting smoking and washing hands.

So Mitchell read the first wave of large European studies about the restorative effects of nature with a great deal of eye rolling. Those studies, published in the early 2000s, linked nearby greenery to everything from longer lives and fewer chronic diseases to higher-birthweight babies. There were simply too many confounds, as he put it. How could any scientist possibly attribute health to nature when the people most likely to be near nature were already healthy, already exercising, already relatively wealthy, and so on? Take Mitchell himself: he grew up tromping around the moors near Exeter in the 1980s with his mum and dad. He read *National Geographic*s in the attic, played bass guitar and enjoyed an early form of geocaching outdoors called letterboxing. His parents suggested he become a scientist, so he did. It would be as preposterous to say it was the windy fens that made him a success as it would be to credit his favorite ham sammies.

Beyond the confounds, "It's easier to understand exercise than nature and trees," he said. The neuroscience is bomb-proof on exercise. Physical activity changes the brain to improve memory and to slow aging; it improves mood and lowers anxiety; in children, it increases the capacity to learn; some studies show it is as effective as antidepressants for alleviating mild depression without the unwanted side effects. By contrast, our collective physical inertia, credited with 1.9 million deaths worldwide annually, is new to our species and getting worse. In preindustrial times, we expended about 1,000 kilocalories per day on physical activity; now we expend an average of about 300.

What changed Mitchell's mind, gradually, was reading the studies from Japan that showed lower stress among forest walkers but not city walkers. There were also some studies showing that people who lived near parks and green areas were healthier, even though they didn't necessarily exercise in them. There was something else going on. And that something else had the potential to make a difference in people's lives.

But he still didn't discount the role of exercise. Time in nature, as the structure of this book suggests, appears to have a dose curve. Five minutes is good; a thirty-minute stroll is better. When you combine exercise and nature, the effects get bigger. "Maybe it's just additive. But maybe it's more than that," he said. To show me, he invited me to join him for some rambling, the favorite national pastime, especially when it involves drinking whiskey.

WE MET AT Mitchell's walk-up garret of an office on campus, out of which he runs the Centre for Research on Environment, Society and Health. Mitchell is wiry and tall, and had to fold himself into his car for the short drive to the edge of town. We'd be ascending Dumgoyne, part of a chain of volcanic hills circling the city to the north. Kitted out with hiking boots, a knapsack filled with "waterproofs" and two walking poles, he eyed my worn sneakers and array of notebooks, cameras and recording equipment. He offered me a pole, but I declined. It was a beautiful day in June, and the countryside was blindingly green. This is one of the most popular day hikes in Glasgow, and I figured the trail would be dry and solid. I'm used to real mountains, after all.

That was my first surprise about rambling in Scotland: there aren't really trails. It's so damp and green that the grass grows faster than human feet can stamp it out. One walks on tufts and clumps of sedges, moss, rock and clover. Straight up, and then straight down.

"This will get your heart rate up," he said. It did, for about an hour. The landscape was ridiculously, lavishly beautiful. We vaulted ancient stone fences lined with blossoming pink foxgloves. Sheep grazed in the fields and a kestrel circled overhead. At the top, we came upon a small group of Boy Scouts. Behind them stretched a 360-degree view of the soft green carpet of Scotland, piling up toward the nearby West Highlands. The color suffused through the land, erasing the roads and houses.

We ate some sammies and took pictures. Before we'd gone very far on the way down, I banana-slipped, scraping my hands but saving the notebook. Mitchell wordlessly offered a pole again, and this time I accepted. I asked him why rambling, as they call it, is so crazy popular in Scotland. ("Hiking," a term reserved for overnight backpacking, is considered a bit of a hippie thing.) Mitchell shrugged and said it's probably because of the country's friendly and ancient right-to-roam laws, which are more lenient than elsewhere in the U.K. and allow you to tromp anywhere across private land, provided you don't steal the sheep, dig up the gardenias, or hunt the landlord's stags. Walking is the most popular sport in Scotland, with Scots taking 2.2 million short walks and 1.8 million long walks per year. I didn't see figures for attendant tick bites, but Mitchell says he digs two or three ticks out of his skin every year.

But it wasn't until we ran into a couple eagerly descending on their way to the Glengoyne distillery that I really understood the national obsession. The hills of Scotland are made of peat, and each region has a slightly different mix of soil, moisture, temperature and exposure. Many of the proper single-malts use barley dried with smoke from the surrounding bog. This is Scottish terroir. We passed a creek, known as a burn, whose water supplies the

Glengoyne distillery before making its way to Loch Lomond, where Rob Roy hid from the English in a cave and, where, nearly a century later, William Wordsworth would fall in love with a dairy maid. To a Scot, each walk is steeped in poetry and spirits, in blood, rebellion and national yearning.

DOWN FROM OUR ramble and back in the garret, Mitchell showed me some bright statistical graphs. In a study that he and colleague Frank Popham published in the *Lancet*, they compared early mortality and disease (in those under age sixty-five) in England with neighborhood green space (defined as "open, undeveloped land with natural vegetation including parks, forests, playing fields and river corridors"). It was a huge study, combing records of 40 million people. "We quite like death as outcome," quipped Mitchell. "We know if they're dead something is wrong with them."

In the greener neighborhoods, death rates were lower for everyone after adjusting for income. Notably, though, deaths were not down for lung cancer, which is not a stress-related cancer and was correctly predicted not to be associated with green space. Cardio deaths, however, were down 4 to 5 percent, which is a big deal given the large population size. But when the researchers looked specifically at death and disease per income level, some interesting patterns emerged. The research showed that income-related health disparities were greatest in areas with the least green. Here, poor people were twice as likely to die as their rich neighbors. In the greenest areas, though, poorer people did relatively much better, starting to catch up to the longer lives of the rich. In other words, there was something protective about the greenery for the most deprived people, either by providing more areas for exercise or by otherwise buffering poverty-related stress.

It's important to issue the standard caveat here; although the study was very large and carefully parsed, it's a cross-sectional study, not a case-control study, meaning it captures a moment in time, making it hard to say with certainty that it was green space and not something else about those neighborhoods causing these effects. So to learn more, Mitchell later analyzed maps, neighborhood services (not just parks but transportation, shops, cultural amenities, and so on) and mental health data from 21,000 residents of 34 European countries, which he published in 2015 in the *American Journal of Preventive Medicine*.

"Only one neighborhood service seemed to have a link with inequalities in mental well-being: green, recreational services," he said. "In fact, inequality in mental well-being among those with the best access to recreational, green areas was about 40 percent less than those with the worst access." This finding would have thrilled Olmsted; the poorest people were the most helped. Parks indeed appeared to be a social leveler. Mitchell has his own phrases for these green spaces: they are "equigenic," and "disruptors of inequality."

But a weird conundrum emerged. When Mitchell turned his attention to Scotland, the pattern wasn't as noticeable. The poorest of the poor were not accessing green space at all, even when it was all around, and Glasgow, as we've seen, is bloody green. Its name means Dear Green Place. But the woodlands near public-housing estates had been neglected, trashed and taken over by ruffians. A favorite park pastime is wheeling in green garbage bins (not the blue ones, they wouldn't do), lighting them on fire and then inhaling the fumes. Not surprisingly, these emerald areas were actually sources of stress. Jane Jacobs anticipated this in her 1961 classic *The Death and Life of Great American Cities*, in which she assailed most city parks as places that "exaggerate the dullness, the danger, the emptiness." Her solution was to throw the baby out altogether, to pave over the parks. Streets and sidewalks, not parks, were the life of the city, she argued. (She was not able to foretell the disappearance of children from sidewalks and the astonishing rise of obesity and chronic diseases.)

Mitchell, on the other hand, saw a failure of civic community. Here was an opportunity for public-health experts to make a difference. And so they are trying. The Scottish government has newly embraced some radical policies. One is cleaning up the woodlands to reinforce medical and mental-health treatment for stressed populations. Another policy, the National Walking Strategy, encourages communities to improve signed trails, organize health walks and otherwise get people off their duffs. It can be a challenging proposition. Consider the scene from *Trainspotting* in which Renton says, "We're colonized by wankers. We couldn't even find a decent race to be colonized by. It's a shite state of affairs to be in, and no amount of fresh air is ever going to change that." But change they'll try.

Government guidelines for the Dear Green Place and beyond state that everyone should have access to safe woodlands within 500 meters of their doorstep. Because for green space to be used, it has to be close. To accomplish this, the country is on a tree-planting and woodland-sprucing-up craze, aiming to increase the percentage of Scotland covered by woodland from 17 to 25 percent. Access to nature is a new national indicator for health in Scotland, and if you squint your eyes and try to imagine the U.S. Congress passing such a standard, you can appreciate just how remarkable this is.

Scotland is so committed to the idea of salvation in the woods, walking or otherwise, that it's underwriting a program called Branching Out to provide mental-health care outdoors. Kevin Lafferty, the health and recreation advisor for Forestry Commission Scotland, invited me to come watch it in action, which is how I came to be molding a clay face onto an oak tree with a group of ex-felons and addicts. The science-based concept is that three hours per week for twelve weeks in a woodland program can reduce symptoms of depression and increase sociability, physical exercise and self-esteem.

Sometimes you meet someone who so easily wears a career, who seems so fulfilled, so unusually capable and perfectly matched to his work that it's clear it's a higher calling. Two such men are Tom Gold and Richard Bolton. Gold works for the Forestry Commission's recreation department, teaching skills like shelter-building to Branching Out participants, and Bolton is a kind of local park ranger, employed by a massive public-housing estate called Cassiltoun outside Glasgow. On the drive to the Cassiltoun woods, Gold kept the windows wide open on the freeway. "Sorry, can't quite get my head around air-conditioning," he'd said as we bombed down the highway.

Tall and wide in a wood-chopping-champion way, Gold had to hunch in the sedan. It was much easier to picture him lumbering through the hills. "My big specialty is bushcraft, the sort of art of making the outdoors a more comfortable place without compromising the resource," he said. "Food, fire, shelter, there are many ways you can achieve or acquire those things, leaving the place exactly as you found it. It's different from survival training, with all the camouflage, traps, gear, weapons and a generally less healthy attitude toward the environment. That's obviously not what we'd do with these guys anyway," he said, referring to the participants, many of whom had recently emerged from institutions. Gold has spent much of his life in the space that intersects mental health and the environment, first working as a leader for a young offenders program in the Arizona wilderness and later in a secure psychiatric facility in Scotland. They were opposite ends of the containment spectrum. In Arizona, he tried to convince the boys that making fire with flint and steel was more reliable than their lighters. "To demonstrate, I inhaled a cigarette and nearly fainted dead on the spot." He saw remarkable changes in the boys, but many returned to gangs once they got back home. "I challenge anyone that age not to get back into it, to resist what all their friends are doing," he said.

In the psychiatric hospital, "nobody was allowed to set foot outside the fence," said Gold. "If it was possible to make a recovery in a nature-based program, that was not on the agenda."

Branching Out, he hopes, can provide both the short-term benefits of a "hoods in the woods" program with the long-term behavioral modifications of more classical therapy. Since its inception in 2007, Branching Out has run some 700 participants through the program, which includes activities such as walking, bushcraft, woodland arts, trail maintenance and birding. The idea is to help people transition from institutions to living more independently. It's been particularly successful in promoting exercise and increasing well-being in the sickest participants.

"We call it ecotherapy," said Gold. "I prefer the term 'adventure therapy,' but it makes some people nervous they'll get eaten to death by mosquitoes while wearing a scratchy wet jumper." Branching Out provides transport, Wellies and waterproofs as needed, and all requisite snacks. It has a long waiting list.

We pulled off the highway and drove up to the old Cassiltoun estate carriage house, where we met ranger Bolton, a small, easy-going man with an air of unhurried competence. He explained that Cassiltoun is home to 13,000 welfare recipients. The unemployment rate here is 39 percent. Drug problems afflict 13 percent of residents and mental-health disorders strike at nearly twice the national average.

But Bolton, who has a background in ecology, thinks these woods can help. He led us some distance into the forest. Although it was sunny and leafy, vestiges of the woodland's delinquent past remained. (In this, the forest is not so different from its users, who retain an air of recent breakage.) I'm not used to seeing tree graffiti, for example. "You should have seen it before," he said. In the three years he's worked here, he's cleared overgrown trails and hauled out 120 tons of trash, including a bus shelter that (along with wheelie bins) people burn to get high. "No wonder they die younger," he said.

To help convey a sense of safety, he often takes classes of schoolchildren here. He's helped organize 108 different cultural and educational events in the past year, led evening health walks and sponsored park worker training. Of the housing residents who have trained with him, 70 percent went on to find permanent employment. He is like Puck: mixing everyone up together in the forest of delights and trusting they'll go back home all sorted out. Like the forest therapists in Korea, Bolton is part naturalist, part social worker, part mythmaker. It's a job description that didn't used to exist, because it didn't need to. We once had a familiar relationship with nature; we knew it on a first-name basis. But now we need professionals to help us reacquaint ourselves with the woods. Soon we may need teachers to remind us how to converse face-to-face. Like a lactation consultant or the people who show us how to bake bread on YouTube, Bolton is a broker in cultural salvage.

At the moment, that meant gargoyles. The small group of depressives, petty criminals and former addicts had assembled on the trail, and Bolton was demonstrating how to make "green men" out of clay and paste them to a tree. The criminal and psychiatric backgrounds of the participants were not revealed to Gold and Bolton. Their job was to work in the present. Bolton kept up an affable monologue as he scurried about. "Along the way I just collected a few wee bits, leaflets; I can start pulling them off and using the shapes, like these sycamore shapes and leaves. Oi! Here's a holly leaf." He was picking them off the ground like a discerning rooster. "The good thing about temporary art, if you don't like it you can start again. You'll notice some of the leaves have quite hairy textures, some smooth. Should I get more color?"

An older man in a yellow windbreaker said, laconically, "Yep." Bolton brushed past a tree dusted with shimmering confetti. "A local nursery uses this as a faerie tree," explained Bolton. "They get a bit heavy-handed with the glitter. This is a lime leaf; it has a nice small point. Woodlands can be your inspiration."

The group gathered around to watch him make a pointy clay nose and fern mustache. Some of the participants looked baleful, some giddy. Their slickers hung loosely and askew against bodies that had gone slack. For many, this would be their first time out of the house all week. But they were obliging. They were six weeks in, halfway; they knew the drill. One man in his early twenties, pudgy with a mohawk, wearing a saggy blue sweatshirt, told me he goes in more for the bush skills than the art. "I like making fires and camping," he told me. He used to do that with his grandfather, when he was a child. He told me he had recently been released from a hospital, that he had scars in the back of his neck. He was glad to be out doing things like a regular bloke. He grabbed a fistful of pine needles and patted them into clay for eyebrows.

Everyone seemed absorbed. It was fun. Making temporary art was a way to be both together and in your own space without high stakes. We admired each other's gargoyles, offering nods and murmurs. The participants, like the gargoyles themselves, represented a wide range of age, color and affect. They were ready for a snack. Gold took over, pulling out an enormous metal pot called a Kelly Kettle. We watched as he demonstrated how to spark a small twiggy fire, first with a bow-like implement out of Sherwood Forest, and when that didn't work, with flint and cotton balls. It was not, let it be said, as speedy as using a Bic. Eventually, he scooped the burning twigs into a ring around the kettle. It boiled the water surprisingly fast. We took tea and biscuits, because that's what Scots do, even in the forest. Many people pulled out cigarettes, because that's what Glaswegians do. They would go home nicely tired, pleased that they'd survived a social outing without any big miscues, looking forward to next week.

For programs like this, the social piece is a large part of it. As Gold put it, "if you're returning to the mainstream after a long period of treatment for mental health, you're not going to go to Queen Street station to see how you get on. You're going to do it in a group where any problems can be examined in a gentle way by people who know only too well where you've been."

BRANCHING OUT IS just the latest incarnation of a long tradition of wilderness-to-build-character enterprises, from the exploits of the seafaring Vikings to Outward Bound. America's best-known outdoor education program, Outward Bound originated in 1939 with a German-Jewish educator and a Briton who had a crazy nostalgia for rough seas. As war was breaking out, they felt young men weren't showing enough toughness, leadership or outdoor training. Great Britain didn't have a lot of wilderness, per se, but it could offer the seas, coastlines and miles of moors. As far as mental-health treatment, Europe had a lineage of psychoanalysis and a tradition of nature-enhanced health spas, so perhaps it was inevitable the two would meet in pastoral hospitals of northern Europe. Interestingly, though, it was an early American psychologist, Benjamin Rush, who first popularized the idea of nature-ish therapy for his mental patients in an 1812 treatise: "It has been remarked, that the maniacs of the male sex in all hospitals, who assist in cutting wood, making fires, and digging in a garden . . . often recover, while persons, whose rank exempts them from performing such services, languish away their lives within the walls of the hospital."

His notions of reform helped slowly change treatment for the mentally ill in America and Europe. Freud had long blamed cities and civilization, at least in part, for unhealthy repressive tendencies. But after World War I, treatment entered a long, mixed interlude of turning mental-health care over to pharmaceuticals, climate control and managed care. To the extent that nature therapy is slowly coming back into vogue, the Swedes have probably done the most to apply science to the field.

The journey of Johan Ottosson seems a good place to start. On a cold winter day twenty-three years ago, Ottosson was riding his bike to work in southern Sweden when he was swiped by a car. He launched many feet through the air, landing headfirst on a rock. He would spend the next six months in a hospital by the North Sea struggling to regain basic skills (he would never read or write without assistance again). It was a miserable, terrifying existence. Although the doctors and therapists were helpful, what Ottosson says really pulled him out of despair and a deep depression were the land and sea nearby.

"I just felt strongly that I wanted to be outside, where I feel the best," he recalled when I went to see him in southern Sweden. "I had a strong relationship with the stones. There is this theory that if a person is in bad shape and low energy, you can't be with other people too much. But you can be with animals, plants, stones and water." Ottosson became so convinced by the healing power of nature that he pursued a doctorate in the topic at the Work Environment, Economy and Environmental Psychology Department at the Swedish University of Agricultural Sciences.

His compelling dissertation includes more details about the span of his recovery, written in the third person. At first, he could only find comfort in rocks. "It was as though the stone spoke to him: 'I have been here forever and will always be here; my entire value lies in my existence and whatever you are or do is of no concern to me.' . . . The feelings calmed him and filled him with harmony. His own situation became less important. The stone had been there long before the first human being had walked past." As he got better, he turned his attention to the ocean waves, and then, gradually, to vegetation, particularly oak trees.

Ottosson's work relies heavily on the mid-twentieth-century American psychologist Howard Searles. Best known for his insights into the idea of transference during psychoanalysis (in which the patient projects feelings onto the therapist), Searles also recognized that nature could provide useful objects of transference. Searles worked at a rural mental hospital in Maryland, where he witnessed this firsthand, writing, "The nonhuman environment, far from being of little or no account to human personality development, constitutes one of the most basically important ingredients of human psychological existence. . . . Over recent decades we have come from dwelling in another world in which the living works of nature either predominated or were near at hand, to dwelling in an environment dominated by a technology which is wondrously powerful and yet nonetheless dead." And that was in 1960.

I visited Ottosson at his campus office in Alnarp. At sixty-three, he has Parkinson's disease and continues to rely on assistants for reading and writing. As he talked, his upper body snaked gently from side to side. He gives talks all over Sweden and is amazed by how many people tell him similar stories of recovery in nature. But it pains him that the modern medical establishment has largely forgotten the insights of Rush and Searles. "When you built a hospital a hundred years ago, you built it around a nice park. That was self-evident. But after about after 1930 or 1940, man is treated like a machine. He gets energy and medicine and that's all. We are just now starting to get fuller knowledge back."

Down the hall from Ottosson in the great historical castle-like building of the landscaping department sits the office of Patrik Grahn, the man responsible for Sweden's nascent renaissance of "horticulture therapy," or using plant cultivation and garden settings as a healing strategy. And the man who inspired him? Ottosson. Grahn wasn't starting from nowhere. As a landscape architect, he'd met the Kaplans in Michigan in the early 1990s, and soon afterward studied the reasons people use city parks in Sweden. He turned up the then-surprising answer of psychological well-being. Then he met Ottosson. "He told me the story about what he experienced and we started some studies. We had wild plans of therapeutic gardens, how they should look," said Grahn, who grew up picking cloudberries and fishing for trout and salmon in Lapland.

With funding from the university, they started a nearby therapy garden complete with a glass-dome greenhouse, water features, flower beds, vegetable fields, pathways and various small structures. Grahn took me by on a gray May afternoon. The first thing that greeted us was a cheerful red garden kitchen skirted by a wide deck overlooking the small fields. Its motto could be the Emerson quote: "The greatest delight which the fields and woods minister, is the suggestion of an occult relation between man and the vegetable. I am not alone and unacknowledged. They nod to me, and I to them." Based on what Grahn learned from Ottosson, the Kaplans and his own empirical research, he believes an effective garden should incorporate a number of elements ranging from safety to fascination to naturalism to species diversity.

It was chilly out and drizzling, so Grahn ushered me into the greenhouse, where therapist Anna-María Pálsdóttir plucked a few leaves off a potted plant and brewed up some citron tea. She explained that Alnarp's standard treatment program runs twelve weeks, like that of Branching Out, but the participants here come four times a week for three hours each time. The Alnarp garden specializes in treating patients with severe work-related stress. They are typically on sick leave, in some cases for years (this being a country that offers sick leave). They are very depressed, lethargic, antisocial and often have other health problems as well. Most take a variety of medications. By the time they get here, "they have cut off everything except trying to stay alive," Pálsdóttir said.

She described the typical progression of patients, and it resembles the experience of Ottosson during his recovery. For the first weeks, the participants often spend their garden hours lying down alone in the garden, either in a hammock or on the ground. Because the program operates year-round, they wear large insulated snowsuits as needed. "Many cannot feel anything" due to severe depression, said Pálsdóttir. "They've almost lost sensory contact from the chin down. As part of healing, the body and the brain connect again. Their interaction with plants trains them to be here and now. They slowly start to pay attention. Things like, what's the tea today, now I can taste coffee and enjoy it. It helps them calm down."

As a former participant—a middle-aged mother named Cecilia who had severe depression—told me later: "I found a hammock near the hedges. It was nice to discover anything outside of the life I'd led before. My brain learned to take in the birds and wind, only that. That's the first thing I remember."

"We point patients to use their senses," said Pálsdóttir. "Eventually, we do creative activities, like go and pick a flower that represents your feelings. Compost what you want to compost. We often use nature as a metaphor that symbolizes good things and bad things. You can stay and be on your own, or help with horticulture if you want. You can just noodle."

"Mindfulness is built in," added Grahn, who between sips of tea pulled out some graphs based on years of published studies. By the end of the program, the patients show a "20 percent drop in symptoms but it's actually more significant than that because the difference is between being considered sick and not sick," he said. According to the World Health Organization, 27 percent of the European population, or 83 million people, experienced at least one mental-health disorder in the past year. If you could speed time to recovery, the savings would be huge. According to Grahn, 60 percent of Alnarp's patients return to work after one year, a figure higher than for those in other kinds of therapy. Based on six years of follow-up data, "the cost-benefit savings is quite high," said Grahn. "They go from seeking primary care thirty times a year to ten." The program is so successful that the Swedish government pays for it and is beginning to replicate it elsewhere. There is a long waiting list to get in.

Grahn is now studying the garden's impact on traumatized Syrian refugees and stroke patients. About 30 percent of Sweden's health-care dollars go to mental health, but stroke care is even more costly. Typically, patients learn to rewire their damaged brains through lots of repetitive speech and occupational therapy, but it's slow and exhausting work. This is where the gardens come in. "There are no established methods of treating mental fatigue," he said, "so we hope we can find a way of treating it for this group. And we hope the environment can help patients find new ways of functioning. A speech therapist takes an apple and says "apple," and shows the object. But in a natural environment, patients can talk and smell and taste and use all the senses, so theoretically it's a more efficient way to facilitate different parts of the brain working together."

THE REASONS THESE programs seem to improve mental and cognitive health is complicated, and by no means is it just about nature and the senses. Nature appears to act directly upon our autonomic systems, calming us, but it also works indirectly, through facilitating social contact and through encouraging exercise and physical movement.

Here's the emerging European coda on public health from Finland, Sweden and Scotland: encourage people—especially distressed populations—to walk, often together, and provide safe, attractive and naturalistic places for them to do it. The research also suggests some special places to go: forests and coastlines. Brits go even more crazy for the coasts than they do for the woods. Basically, the closer you live to the ocean, the happier you are. Researchers at the University of Essex School of Health and Human Sciences found that if you live near the scenic western coasts of England, you're nine times more likely to exercise than other people, even after adjusting for income. As the epidemiologist Ian Alcock put it, if you want to be happy, there is a simple, scientific formula: "get married, get a job and live near the coast."

Parsing the research apart further, if you are depressed or anxious, social walking in nature boosts your mood, assuming you're walking with people you like; if you want to solve problems in your life, self-reflect and jolt your creativity, it's better to go alone, in a safe place.

I find myself most drawn to the fate of the solo walker, because I tend to be one. I love a good hike with a friend, but I consider it a gabfest more than anything. I'm protective of my solo walking time precisely because I have found it to be so helpful in solving problems, personal and otherwise. What is it about that peculiar synergy of walking plus nature? Being in Scotland got me thinking about Wordsworth, creativity and the essence of imagination. Walking lies at the heart of it. Although these topics remain somewhat mysterious to neuroscientists, the poets can offer some assistance.

Rambling On

When we walk, we naturally go to the fields and woods: what
would become of us, if we walked only in a garden or a mall?

—HENRY DAVID THOREAU

The idea of *solvitur ambulando* (in walking it will be solved)
has been around since St. Augustine, but well before that
Aristotle thought and taught while walking the open-air
parapets of the Lyceum. It has long been believed that walking in
restorative settings could lead not only to physical vigor but to mental clarity and even bursts of genius, inspiration (with its etymology
in breathing) and overall sanity. As French academic Frederic Gros
writes in *A Philosophy of Walking*, it's simply "the best way to go
more slowly than any other method that has ever been found." Jef-
ferson walked to clear his mind, while Thoreau and Nietzsche, like
Aristotle, walked to think. "All truly great thoughts are conceived
while walking," wrote Nietzsche in *Twilight of the Idols*. And Rousseau wrote in *Confessions*, "I can only meditate when I am walking.
When I stop, I cease to think; my mind only works with my legs."

Scotland clearly relishes its twin legacy of brains and long-striding.
On the wall of the National Museum of Scotland hangs a quote from

James Watt, inventor of the steam engine (yes, the steam engine) in 1765: "It was in the Green of Glasgow . . . when the idea came into my mind, that as steam was an elastic body it would rush into a vacuum. . . . I had not walked further than the Golf-house when the whole thing was arranged in my mind." Nikola Tesla, too, invented a revolutionary engine while on a long walk in a Budapest park. Little did these men know how transport engines would hasten the demise of pedestrian life.

Anticipating the exercise/nature debate, Thoreau opined, ". . . the walking of which I speak has nothing in it akin to taking exercise . . . but is itself the enterprise and adventure of the day." He also wrote, in his essay "Walking," "I think that I cannot preserve my health and spirits, unless I spend four hours a day at least—and it is commonly more than that—sauntering through the woods and over the hills and fields, absolutely free from all worldly engagements."

Walt Whitman was an even stronger evangelist on the topic, exhorting men to be more perfect and more manly by striding around outside. "To you, clerk, literary man, sedentary person, man of fortune, idler, the same advice," he wrote. "Up! The world (perhaps you now look upon it with pallid and disgusted eyes) is full of zest and beauty for you, if you approach it in the right spirit! Out in the morning!"

If for them nature provided mental clarity and adventure, for Wordsworth it provided sanity itself. Nature, as he declared in "Tintern Abbey," was "the nurse, / The guide, the guardian of my heart."

It's worth taking a short perambulation to the poet's sensibility, not just because he was the Romantic Age's greatest advertisement for both Scotland and for perambulating (he is estimated to have walked some 180,000 miles in his lifetime, composing poems as he went), but because he wrote so often about the ways in which his own mental health was bound to nature, and he was the first to do so in a thoroughly modern voice. Dismissing Wordsworth as a daffodil-

gazing nature poet would be a mistake. His greatest defender of
recent times has been the late Yale scholar Geoffrey Hartman, who
argued that Wordsworth essentially invented modern poetry (with
a small assist from Coleridge), and in so doing saved the art form
altogether. I'm fascinated by how Wordsworth intuited the neuroscience in both psychology and cognition. We forget today that poets
were the philosophers of their time, and that the good ones changed
the course of history.

Wordsworth was a child of trauma. His mother died when he was
eight and his father when he was thirteen. He was sent off to live
with unsympathetic relatives. Money was tight and the siblings
lived apart. It's hard to overstate the stress of these events, and
at a critical time in the development of the poet's psyche. Hartman's own history followed a similar trajectory. In 1939, at the age
of nine, he and dozens of other boys were plucked from a Jewish
school in Frankfurt and sent to live in an outbuilding on a country
estate in England. He remained there for six years until the war was
over, when he was finally able to reunite with his destitute mother
in New York.

Hartman celebrated and summarized one of Wordsworth's central themes: "Nature does everything to prepare you, to make you
immune, or to gentle the shock. He doesn't say there is no shock, or
surprise, but that nature aims at a growth of the mind which can
absorb or overcome shock."

A few months before Hartman died in 2016, I called him up. In
his mid-eighties, he was still living in New Haven. I had taken a
class with him in Romantic poetry at Yale more than two decades
before. I wanted to see if he could once again help me through some
of the material. Mostly, though, he wanted to talk about what Wordsworth meant to him all those lonely years ago, during his own
period of shock. "I think the comfort of nature and the comfort of
enjoying poetry and being encouraged to read, including especially

Wordsworth, certainly helped to make my exile a little bit more tolerable," he explained. "I hadn't enjoyed nature before England. . . . So going to England and reading Wordsworth reversed my sense of things." Perhaps it was inevitable that Hartman would be the one to rehabilitate Wordsworth's reputation in postwar academe.

As Hartman reminded me, Wordsworth made the perceiving self central to perception. Nature was meaningful precisely because of how it "interfused" with the mind, forming the basis for imagination. This is a central theme in the first book of *The Recluse*, a long autobiographic poem written in 1798. "How exquisitely the individual Mind/. . . to the external World/Is fitted:—and how exquisitely, too—/. . . The external World is fitted to the Mind." And sitting on the banks of the River Wye, the poet marveled at how "an eye made quiet by the power / Of harmony" offered relief from "the fever of the world." Nature had certainly offered that relief to Hartman, and I imagine it may have in his final months as well.

Wordsworth is sometimes credited with launching the idea of tourism, but at least equal credit should go to his sister, Dorothy, who slogged many, many miles with him and wrote *Recollections of a Tour Made in Scotland* in 1803. It's a great read, not only because it depicts Coleridge as wet and cranky, but because it recounts things like eating boiled sheep's head with its hair singed off. Wrote Dorothy Wordsworth: "Scotland is the country above all others that I have seen, in which a man of imagination may carve out his own pleasures. There are so many inhabited solitudes, and the employments of the people are so immediately connected with the places where you find them."

Both siblings were inveterate Romantics, reacting against the march of industry and commerce into pastoral landscapes. While cities had once offered excitement and revolutionary ideas to a young William, he later came to believe that they embodied disillusionment and stagnation, a "savage torpor." Far from making people more creative, the din and grime stifled their dreams, or at least his.

The Wordsworths were contemporaries of Jane Austen, whose *Pride and Prejudice* appeared in 1813. The notion of walking as an expression of good breeding and good health was in full swing, but it also enabled an outlet of independence rare for a woman, and both Dorothy Wordsworth and Austen's heroines relished the act. As the essayist Rebecca Solnit points out in *Wanderlust: A History of Walking*, when Elizabeth Bennet charges out alone across the muddy downs to help her ailing sister at Darcy's place, she is rendered both slightly scandalous and alluring.

By the early nineteenth century, it had become hard to disentangle walking and its hale enthusiasts from the Enlightenment, from Romanticism and, thanks to Thoreau and Emerson, from budding American nationalism. Walking was a philosophical act, facilitating a direct experience with divinity. It was a political act, mixing the educated classes up with the poor (who had always walked, doh). And it was an intellectual act, generating ideas and art. The ramblers of yore embraced a kind of radical common sense.

Today, when everyone from corporate executives to distracted "knowledge workers" are obsessed with creativity, walking is getting a new look. Executives hold walking meetings and even walk on treadmills at their desks (a terrible idea—go outside for a real walk!). People everywhere obsess over their step-counting wearable devices. They organize community walks. And if they are the sort of scientist I've been writing about in this book, they also walk with portable EEG units—or make their subjects, and inquisitive visitors like me, go out and do it for them.

THE ABILITY TO see electrical waves inside the human brain was pioneered by German psychiatrist Hans Berger in the 1920s. Berger, who fell off a horse as a young soldier and was convinced his brain then sent a telepathic message to his sister, wanted to investigate. He also believed it should be possible to watch the brain convert energy into blood flow, electricity and, ultimately, thoughts themselves. What started off as a kooky quest eventually led him to invent the electroencephalography machine, which translated signals from electrodes placed on the head to a photographic recording device. He referred to the contraption as a brain mirror, although that was optimistic. It wasn't able to read or reflect minds but it could capture electrical signals that revealed clues about mental states. Berger learned that alpha waves, for example, appeared during rest or relaxation. Later, there would be other insights, such as that beta waves indicate active thinking and alertness, that gammas dominate during sensory processing, that delta occurs in deep sleep and so on.

Until recently, EEG was complicated to administer, requiring tight skullcaps fitted with dozens of button-sized electrodes, each wired to a large computer. A person wearing such a device looks like a shriveled sea urchin. But now, thanks to wireless technology and microprocessors, subjects can take those electrodes for a walk, as long as they don't throw their heads back and forth in abandon (for this reason, we have no idea what the brain looks like while dancing). Although EEG remains a relatively crude measure of the average electrical output of thousands of neurons over a wide area of brain geography, it holds an obvious allure for researchers interested in environmental psychology.

In a small but intriguing 2013 pilot study, researchers asked a dozen volunteers to walk around Edinburgh for a total of 25 minutes. Their path took them through a busy urban thoroughfare, a city park, and a quiet street. The walkers wore a newfangled portable EEG that wraps just a few plastic tentacles around one's head, made by the California company EMOTIV. The unit has only 14 electrodes and transmits real-time information wirelessly to a laptop. EMOTIV then runs the frequency signals of alpha, beta, delta and theta waves

through an algorithm that translates them to short-term excitement, frustration, "engagement," "arousal" and "meditation level."
(This is also the same kind of unit I wore on the lake in Maine.)

When the Scottish volunteers entered the park, their brain waves
showed evidence of lower frustration and arousal, along with higher
"meditation" levels. Encouraged that these results aligned with
Attention Restoration Theory, the researchers have now launched a
much larger study with 120 senior citizens. They are calling it the
Mobility, Mood and Place study.

The lead researcher, Jenny Roe from the University of York, agreed
to let me have a go with the EEG unit on the route through Edinburgh.
I met her neuroscience postdoc, Christopher Neale, downtown,
and after a bit of hair maneuvering and saline-solution dabbing, he
clamped on the headset. "You have a lot of hair," he muttered. "That's
one difference about working with older people. They're mostly bald."
But the device was finally transmitting, and so with Neale leading the
way about ten paces in front of me, we began the walk.

It was a beautiful June day. We headed down Chalmers Street,
bustling and loud with students, lorries, buses and motorbikes. This
was gratifying, because I knew the noise would stress me out, and
of course I knew the study design (which does not make me an ideal
subject). Then we turned into the Meadows park, and I prepared to
calm down. But I couldn't. The park was jam-packed with picnickers, baby carriages, joggers. Boom boxes blared from the picnic
blankets. A park maintenance truck was backing up out of a small
dirt alley. Oh no! You people are all messing with my solitude! This
is generally my attitude while in city parks, but it was exacerbated
by the pressure to produce good brain waves. Look at the grass, I
willed myself. Listen to the damn birds. A bicyclist careened past.
We exited the park and walked up a quieter street, ending up near
the National Museum. Neale unclenched the unit from my now
throbbing head and promised to send me the results.

Months later, I got the analysis of my brain waves back from Neale. It was a bit disappointing, if not surprising. "You can see that when you transition into the green space, your excitement, engagement and frustration levels all go up," he wrote. "These results suggest that you were more excited and engaged in the green space when compared with the urban busy section. Interestingly, your frustration levels go up and remain up. Perhaps this was due to the fact that you were walking around a new city, and technically 'at work' too!"

More likely, I was just, like Wordsworth, pissed off by the crowds.

In any case, I was, as Neale put it, "non-typical. Early results using the raw EEG data in our newer study in older people are promising and more in line with our hypothesis, i.e., that walking in a green setting is restorative." Something Ruth Ann Atchley said in Moab came back to me, about how she thinks different people have different tolerances for doses of "nature." Someone who lives in a city might be overjoyed and calmed down by a single tree, but others of us require a bigger hit. "If you're used to Colorado, you're going to want quiet and big views," she'd predicted. Nature was like caffeine, or heroin. You keep wanting more.

I was, it seems, spoiled.

OR I COULD just be a terrible research subject. A few months later, I traveled to Urbana, Illinois. I went to visit Art Kramer, the exercise neuroscientist, rock climber and Harley rider whom I'd last seen fidgeting on a deck chair in Moab. It was apparent he didn't like to sit still then, and when I saw the sixty-three-year-old's office at the University of Illinois' Beckman Institute for Advanced Science and Technology, it was even more obvious. As the institute's director, he commanded a wood-paneled office large enough to accommodate a treadmill desk.

"One to one and a half hours per day," he said, as I sized it up.

THE NATURE FIX 177

"One point seven to two miles per hour." Kramer, who has expressive, sunken eyes, a trim gray beard, and an appearance of explosive energy modulated by sensitivity, was wearing a slightly rumpled striped shirt, and I wondered if he had just climbed off the thing.

Kramer has made many academic splashes, but a big one was when he figured out that forty minutes of moderate walking per day could protect the aging brain from some cognitive decline, especially in executive function skills, memory and psychomotor speed. To exercise, he has added a list of additional advice: have good genes, stay intellectually challenged, maintain social interactions. He has even advocated walking book clubs, which, I must say, sounds not nearly as fun as curling up on couches with dessert and glasses of wine. And thanks to his colleague and former student David Strayer, he's taking a look at nature as a way to boost creativity. After attending Strayer's desert confab, "I thought looking at nature would be a great idea," he said. "We can begin to look at the synergistic effects of nature and exercise. We can try to isolate it in a lab."

Kramer was intrigued by a recent Stanford study that showed walking on a treadmill and walking outside both increased divergent creativity, which is the kind of expansive thinking that includes brainstorming and finding more than one correct answer to a question. That study did not show that walking improved convergent creativity, the kind exemplified by the word-association task that Strayer used showing big payoffs in Outdoor Bound hikers (as a reminder of the task, find the one word that connects to all three words: cake, cottage and Swiss—the answer, in case you're not hungry enough to free-associate it, is cheese). But the Stanford study did not look at walking in nature per se. The "outdoor" part took place on campus streets, alleys and courtyards. Stanford may be beautiful, but it is also loud with people and service vehicles, as I learned when I walked the route myself. Naturally, it was during a walking meeting that Stanford professor Daniel Schwartz and his

Ph.D. student, Marily Oppezzo, got the idea to study walking and creativity. Because they were being so dang creative on that walk.

Wanting to work in the nature piece, Kramer thought he'd dish out a few creativity tasks before and after putting volunteers on a treadmill for twenty minutes. Some would "journey" through a virtual-reality park, and some a city street. Of course, I wanted to try it. Kramer's grad student set me up. From the get-go it was a disaster. The pretest was to create a list in a category, in this case "animals," coming up with as many as you can in a set amount of time. I was on a roll, probably because I once lived on a game ranch in Africa. I was up to wildebeest, oryx, black rhino and water buffalo when the timer buzzed. This was a problem. In order to show that nature makes you more creative, you're not supposed to ace the pretest.

It was time to mount the machine. The treadmill faced two enormous screens running the 3-D video of the walks. I started ambling at a comfortable pace, but the machine made a loud whirring noise in the windowless room. This did not feel like a pleasant nature environment. Not at all. The room was stuffy, the machines loud, the images on the medium-pixelated TVs glaring. VR, I was learning, is much more V than R. When I shifted my gaze from the left screen to the right, the picture quality there was so bad that the trees looked like they had been dusted with nuclear ash. Then a bright flash would burst and the image would shake and reset. I felt woozy, as I had the last time I'd gone virtual in a lab. I waved down the assistant, who managed to switch the video to 2D before I felt the need to hurl. Afterward, I took the word-associates test.

I bombed.

But, apparently, so did other people. Kramer told me later the study "was a bit of a bust." There were problems with the lab technology, specifically the "presentation of scenes across multiple screens and mismatching auditory and visual scene elements." Perhaps it's time to admit it, people: nature just does the elements better.

DAVID STRAYER HAD been having better luck with his post-Moab experiments than Kramer. He conducted his own walking study outside, per his style. "We know the field is messy," he told me. "There's wind and rain. But being in the lab strips away a lot of the interesting stuff, so I've learned to grin and bear it and accept the consequences."

Strayer decided to make use of the Red Butte arboretum near the University of Utah campus. He wanted to look at the effects of being in nature on walkers' memory, and he also—because he is David Strayer, Distracted Driving Man—wanted to look at how technology use might mess with memory. For the experiment, Strayer and doctoral student Rachel Hopman set up three groups of about twenty people each: one group would hand over their cell phones, walk for thirty minutes in the arboretum and then take a recognition memory task. A second group would take the same walk and test, but during the walk, they were told to make a long phone call. Moms were happy that day. The third group was the control. They took the memory test before the walk. The first group, walking with no phone, averaged 80 percent in their postwalk memory test. The group that talked on the phone scored only 30 percent, and the control group scored about the same.

Strayer was delighted to see both that nature walking boosted cognition and that the addition of evil technology totally wiped out the gains. "What we find is consistent with the other literature that working memory improves," he said. And, he explained, it is also consistent with the Kaplans' Attention Restoration Theory. The quiet hikers were able to access the Kaplans' magic recipe of feeling "away," of being open to soft fascination in their environment, of having a sense of compatibility with the landscape and feeling as if they are in a vast, restful space. The phone talkers, by contrast, may have been relaxed by being outside in the fresh air, but they were not as liberated from daily cares. They weren't truly resting their top-down attentional networks. They were multitasking, walking, looking, listening and most importantly, *speaking*, which uses up a lot of attentional bandwidth. Note to self: leave the cell phone at home, or at least deep in your pocket, when in need of a cognitive reboot.

About the same time Strayer was running his experiment, yet another Stanford team designed a walking-in-nature study (it's interesting to note that the campus most known for changing our relationship to technology—by incubating it—is now becoming known for helping us ditch it). As sometimes happens, neither team was familiar with the other's work, but there was some nice complementarity. Working with ecosystem services expert Gretchen Daily and emotional-regulation-psych guru James Gross, doctoral student Greg Bratman randomly sent sixty volunteers on either a fifty-minute walk through a busy street in Palo Alto or on trails around the iconic local green space known as the Stanford Dish. Before and after, he measured their moods, anxiety and rumination, and also gave them a series of punishing cognitive tests. Results? The subjects performed significantly better on a test measuring memory and attention—and they also reported feeling happier—after walking in nature.

Bratman and his colleagues had a theory about why, and they wanted to test it. His coinvestigator, Gross, is an expert on rumination. This is something cows do literally, but our minds do it too: chew on an unpleasant memory to create, as the study authors put it, "a maladaptive pattern of self-referential thought." We might replay an unpleasant exchange or bad feeling over and over until we drive ourselves batty. Rumination, as Gross and others have shown, is linked to depression and anxiety. When people ruminate, they activate a portion of their brains called the subgenual prefrontal cortex, a region also linked to sadness, withdrawal and general grumpiness, according to Bratman.

For the next experiment, they sent 38 healthy (not depressed) city dwellers on a pretty big walk—90 minutes this time—either back to the green Dish or along traffic-heavy El Camino Real, and scanned their brains before and after. They also had them fill out rumination-measuring questionnaires. On the scans, the nature brains showed a significant, sizable reduction of blood flow to the subgenual region, while the urban brains showed none. The questionnaires also revealed less broody feelings in the postwalk Dishers but not in the roadway walkers. The results were exciting for Bratman, because they point to a possible causal mechanism for *how* certain landscapes might be boosting our moods, basically, by quieting some brain circuitry governing self-wallowing. The world is bigger than you, nature says. Get over yourself. At the very least, nature distracts us the way a parent might distract a whining toddler, by waving a favorite stuffed animal. As Bratman put it, "The results suggest that nature experience is impacting rumination in a way that is markedly different from urban experience."

CLEARLY, IT WAS time for me to get walking. I was, despite trying for nearly two years, still feeling unhappy in D.C. The city sounds jangled me. We were hemorrhaging our savings. My husband had a fulfilling job saving nature, but we had to leave wild landscapes for him to do it, which still rankled. What about saving us? I was grateful to spend more time with my father, who continued his impressive recovery from his brain trauma. Together, we took increasingly longer walks in an arboretum near his place or along the canal near mine. He was happier and mellower after his accident, and, walking, he often brought up pleasant reminiscences (as opposed to ruminescences) and some pretty sappy sentiments. I haven't seen any studies on nature and sentimentality (hear that, Bratman?), but the connection wouldn't surprise me. One day, as we returned to my front steps, Dad thanked me. "You are the light of my life," he said.

"Wait a minute!" protested his wife, Galina. He laughed.

"You both are." We had a group hug, reminding me that nature is, truly, best shared.

To motivate myself to get out walking more, I found a study I could join, a big, old-fashioned study with questionnaires.

I learned that Lisa Nisbet at Trent University was sending over 9,000 people out into the verdure for the May-long "30 x 30 nature challenge"—30 minutes a day of walking, for 30 days in a row). I signed on. My first task was to answer a fairly long questionnaire designed to ascertain our general mood state, vitality, activities and "subjective connection with nature." That done, I set out for my walks, generally down to my usual path along the C & O Canal, but in one case along a park in the late evening in downtown Helsinki, where a man stood in a clearing and waved his penis around.

When we are determined to hobnob with greenery every day, most of us will, inevitably, encounter setbacks. Over the course of writing this book, I was jumped by numerous rogue and grimy dogs and splattered with mud by bicyclists. I broke a finger when my own dog lunged for another dog on a crowded park trail, wrenching her leash around my hand. I was stung by four bees, three in D.C. One morning I was seized by an unstoppable urge to go to the bathroom and hurriedly plunged into the dark creekside thickets of my neighborhood park (please don't tell the listserve). I consequently contracted poison ivy. The Lyme disease came later, from Maine.

It's not easy being outside everyday. Either a lot of people in Nesbit's study decided they preferred the air-conditioning, or they simply didn't respond to the follow-up questionnaire. Of the 2,500 who stuck it out, most were just like me: women in their mid-forties. Researchers love us because we do, sigh, follow through on our commitments, and we are conditioned to be helpful. But there were rewards: I spoke to Nisbet months later, after she'd sorted the data. "The more time participants spent in nature, the greater well-being they reported," she said. One of the most interesting findings was that we seemed to like being in nature so much, we doubled our weekly green time by the end of the month, from five hours to ten. As the month progressed, we also reduced our time in vehicles, texting and emailing. Progress! All this temporal rearrangement appeared to be good for us. We reported significant increases in all measures of well-being, including in mood and mental calm, and also decreases in stress and negativity. We slept slightly better, and also reported feeling slightly more connected to nature.

This was all true for me. The more I made myself get outside, the better I slept and felt, except when my bee-stung arm turned into armzilla. But the discomfort was temporary. Despite the planes and all the people, my nearby parks were invariably cooler, breezier and better-smelling than anywhere else in the city. I watched the buds turn to leaves and I made a point of trying to identify a few birds by sound and of looking for fractals. I often walked to look at the Potomac River, just to take the currents in and let the water (always the highest-rated nature feature in surveys) work its magic on my tired neurons. The required thirty minutes often turned into many more.

Still, it felt a little contrived. Pull out the stopwatch. Try to feel connected. I wanted to find people who were spending even more intensive time in nature, real nature, and, frankly, I wanted it myself, now that I was all connected.

It was time to head for the backcountry.

PART FOUR

BACKCOUNTRY BRAIN

Get Over Yourself:
Wilderness, Creativity and
the Power of Awe

Calvin: Look at all the stars! The universe just goes out
forever and ever!
Hobbes: It kind of makes you wonder why man considers himself
such a big screaming deal.

—BILL WATTERSON

David Strayer never gets tired of watching his college students tumble down the wilderness whirlpool into a new head space. Every April, he takes his advanced psych class, called "Cognition in the Wild," to the desert for a few days of camping, exploration and yes, a mental boost. Phone use is vigorously discouraged, not surprisingly. Billing it as a seminar on how our mental experience is connected to the environment, he's been teaching it at the University of Utah for eight years. The annual field trip is part of what's driven him to pursue his "three-day effect" theory, of senses, perspective and cognition sharpening over time. This year, he invited me to see it unfold and to try out his latest experiment building off last year's Moab confab.

Just before dark, I pulled into the Sand Island campground along the San Juan River near the tiny, dusty town of Bluff, Utah. Strayer was serving fajitas out of fire-blackened pots. It was 36 degrees out, and that afternoon the students had driven down through a

foot of new snow around Salt Lake City in what the radio was calling the Tax Day Storm. Now a group of about thirty undergraduates
and research assistants packed in around the campfire, scooping
up their hot food with gusto. One student was pouring Sprite into
the dessert pan for peach cobbler, college style. It would taste like
an explosion of sugar. When the stars came out and the hot chocolate was poured, Strayer announced it was time to start the nightly
round of ten-minute research presentations on topics like urban
stressors on athletes and teen cell-phone use (teacher's pet!). I
pulled on my gloves and settled in. For the students, participating
on this trip would encompass 30 percent of their grade. Strayer, who
was, naturally, a Scoutmaster when his boys were young, said he
believed the campfire setting was vastly superior to power points in
classroom. "Here, they really raise their game," he told me. "By fire
they come alive."

He's not the first to think so. The French philosopher Gaston
Bachelard wrote in 1938 that fire "begat philosophy." In drawing
us together for meal preparation and warmth, fire drove evolution,
selecting those of us who could be sociable, communal and even
entertaining. We needed the warmth on this night, and I marveled
how unusual it was to see a group of young people looking at one
another or gazing into the lumens of the fire and not into the lumens
of their phones.

The next morning, after a thoroughly disreputable breakfast of
Pop-Tarts, muffins and strawberry yogurt from Costco, we drove off
to an unmarked trailhead along Comb Ridge. This eighty-mile-long
monocline rises from the desert floor, gouged along its east side by
deep gullies and canyons that were once home to the Anasazi people. Although they vanished eight hundred years ago under mysterious circumstances (most likely drought and war), many of their
artifacts, wall art panels, and rough stone dwellings survived well
in the arid desolation.

Strayer led the way up a sandy trail that soon hardened into solid rock marked by cairns. The day warmed and we tied our layers around our waists. One young woman in a ponytail wore red shorts with the word UTAH written across her seat. Some students bounded ahead comparing notes on the latest Michael Keaton movie and some straggled behind, unused to exertion. Overall, the vibe of the class was less jocky, more nerdy, wearing less high-tech clothing and more nose rings and blue nail polish than I expected. For many, this was their first time in canyon country. Most of them didn't know each other outside of class.

Before long we came to a half-crumbled dwelling nestled into a smooth concavity in the cliff. Pottery sherds lay about, and you could still make out the rounded rooms of the ceremonial kivas. Faint red handprints and human-figure drawings frescoed the cave's back wall. The place had been hastily abandoned in desperate times. It was eerily quiet among these ancient bedrooms and prayer rooms. We continued farther on toward the crest of an exposed ridge and a breathtaking frieze known as Procession Panel. Believed to date from the "basket-maker period" around 700 A.D., it depicts a tight line of figures migrating from some sort of portal, either spiritual or literal. It presides along an ancient trail connecting two parts of the Anasazi realm.

Over the following days, we ambled around similar sites, from a vast wall painted by one artist known as "wolf man" featuring ducks, yucca plants and what might be human heads shaped like light bulbs, to ruins with names like Split Level and Long Finger. Our senses of perception were shifting. The faint scrapings of rock art that at first were hard to discern started popping out. We could spot the smooth stones used for grinding, the sharp bits that were broken pots. Strayer would point to a thousand-year-old corn cob or examine some pottery and declare it from a certain period based on the clay and firing technique. During an alfresco lunch, he described how one clan held a monopoly on a recipe for oxidizing clay to make it red, guarding the secret and prospering in trade.

"Technology is always a double-edged sword," said Strayer, fingering a delicately corrugated sherd before passing it around. "It enabled progress but it changed who they were. The cowboys who dug up bones here suddenly starting finding small skulls with flat heads. When the people here started cultivating corn, the mothers had to tend the fields, and they swaddled the babies' heads flat against a carryall. The evolution of technology is who we are, the stepping stone, with inventions embodying new ways of thinking and being from which we can't go back." He seamlessly segued to his own burdens with technology. "I'm sure when I get back I'll have three or four hundred emails. Most of them will no longer mean anything."

If Strayer wanted to wow them, he was succeeding. Most of the students seemed impressed, even amazed, by these remote finds and dramatic rock fissures. "I didn't know I was going to be deeply affected by this," said Lauren in pink sunglasses, her black hair in a messy bun, "like when I saw that handprint, I almost cried. It's so unlike me."

Heading out on morning three, we were met on the trail by a great horned owl that sat still as a statue on a stone ledge over our heads. Amelia, a blonde with a sorority vibe, squealed, "I've never seen one before!" Earlier, she had admitted to her tentmates that she was missing her phone because she was waiting for a cute boy to text her. But now, she was transported. "You guys! I feel like I haven't lived until this trip!"

We lunched in clumps among the blooming prickly pear where Butler Wash meets the wide, gently flowing San Juan River. At our backs loomed a sheer, smooth golden wall; to the south and west lay an expansive spread of the river and its surrounding upheaval of multicolored sandstone. Strayer told us about a petroglyph panel some ways downstream, accessible only by wading and swimming and then returning against the current. It was finally warm out, and a handful of students decided to pursue the lead. They wouldn't return to camp until early evening, flushed with adventure, giddy, triumphant and hungry for Strayer's hearty cooking. They had made their own grateful procession through the raw, spare, sometimes voluptuous country.

Strayer was delighted the students were exercising their exploratory—and social—instincts. "The students have gelled," he told me on the way back to camp. "It just shows you how starved they were for social interaction, for connection." I had to wonder if he was projecting his usual technology-has-ruined-young-people bias, but the fraying of social skills is increasingly documented by researchers such as Sherry Turkle at the Massachusetts Institute of Technology. Our capacities for empathy and self-reflection do appear to be challenged—even atrophying—as our digital interactions replace analog ones. One happy solution Turkle acknowledges but doesn't emphasize: spending more time in unwired places. One of the underappreciated benefits of venturing into remote landscapes is that we are often thrown into connecting with each other.

Just before the adventurers returned, it was my turn to undergo Strayer's latest experiment. His grad student Rachel Hopman tucked my head into an EEG device more elaborate than the crown-of-thorns one I wore in Scotland and on the lake in Maine. It was more like a bathing cap with twelve sensors sprouting out. Six more sensors suckled my face, all connected via many wires to a small portable unit beside me. I felt like a tethered hedgehog. I carefully settled into a lawn chair at the tamarisk-lined edge of the campground along the San Juan. The students and I would be sitting here in pairs for about fifteen minutes, not doing anything in particular. Different groups of subjects would be doing a similar nothing while sitting at the edge of a parking lot in Salt Lake City and in a lab with a computer.

This was all an elaborate field experiment that grew out of the previous year's Moab gathering. Strayer wanted to find a biomarker that could show a brain under the influence of nature. If, as most seemed to agree, something is happening to our brains, is there some way to see the transformation? Adam Gazzaley, our rooftop margarita maker from the University of California, San Francisco, had lured Strayer with the idea of measuring midline frontal theta waves. Because these brain waves increase in power when the frontal cortex is engaged in an executive task, Strayer and Gazzaley were hoping the opposite would be true during a wilderness mind-blow: the thetas would quiet down, potentially indicating a rousing of the dreamy default network instead.

If a river can't transfix my brain, then nothing can. I've spent a lot of time in this book talking about trees, but when I crave wild places, it's often the desert I want. In his wilderness-defense classic *Desert Solitaire*, Edward Abbey named a chapter "Bedrock and Paradox" after towns not far from here. It's the perfect nomenclature for a landscape that is chaotic and static at the same time, the rock as dry as a cow skull but broken by lush shocks of green. In the aridity, the greens are greener and the blues are bluer, and, as Abbey puts it, "all things are in motion, all is in process, nothing abides, nothing will ever change in this eternal moment." Ellen Meloy, an essayist more subtle and interior than Abbey and who lived and died near Bluff, remarked that this county was the size of Belize and contained not one traffic light. "The nights are coalblack and water-deep, the light often too bright to understand. . . . No one is ever sure if we are hostages of isolation or the freest people in four states."

Of course, the ultimate paradox is that humans need both wilderness and civilization, and that one makes us all the more poised for the other. Although I grew up in New York City, I dreamed of wild summer landscapes unfurling before me. They lay loosely threaded together by the rivers my Dad and I ran, including this very one, launching from this very campground twenty-nine years earlier.

The main watery artery of this region, the San Juan River seeps and then gushes out of Colorado's southwestern mountains, joining the Colorado River some 380 miles farther down. At that point it is technically no longer a river but a giant, placid lake created by Glen Canyon Dam. Like us, the river fully transforms from wild to domesticated, but it has no option for reversal. Packed into my EEG cap, I watched the river as delicate fractal patterns of flow played against the light. The milky chai-colored water rippled and coursed, shallow in sections, braiding along its main channel.

Sitting here, I felt washed over by the calm of the scene, but it was also mixed with a little anxiety about another weather system approaching from the west. We had no cell reception to check our weather apps. Anxiety may thrive in cities, but it's also at home in the wilderness—another paradox.

LATER, WHILE HIS enchilada pies were baking in their cast-iron Dutch ovens, I asked Strayer what he thought of the fractal/visual theory of brain restoration, the idea that when our visual cortex finds a sweet spot of information, it can trigger our pleasure centers and help relax us. He wasn't overly enthused. What he's getting at, he explained, is a change in mind-set that occurs over hours and days. The kind he and his students have just experienced, with their mild sunburns, loosened limbs, easy laughter and fresh insights.

"If it's just the visual cortex," he asked, "why can't I watch *National Geographic* videos and get this sensation? I don't feel this and I couldn't watch four days of it, and those are amazing videos."

"But a few minutes out a window can improve your mood and drop your blood pressure," I said, citing studies as Strayer lifted a heavy lid to check on dinner.

"What I'm interested in isn't that. That's not what I and Abbey and Muir and Thoreau are talking about. It's something much deeper, more cutting close to our soul. Frankly, it's the essence of who we are and getting away from the rat race, across the litany of literature."

Satisfied with the progress of cheese meltage on his enchiladas, he pulled off his oven mitt. "If I was a betting man, I'd be betting on the fact that the prefrontal cortex is not in overload in nature."

STRAYER IS A betting man, because he was out here spending a pile of the National Academy of Sciences' money on EEG machines. It seemed to me that when the brain is "resting" from its onslaught of daily tasks, it's making room for something else. It might be the default network—the one that spurs daydreams and reflection—but it might not be. One conundrum is that the most accomplished Buddhist meditators, the ones who've spent tens of thousands of hours mastering that prized calm-alert state, don't appear to be firing up their default networks when they meditate. What they're accessing is something not easily mapped in discreet places in the brain, but the circuits seem to be related to feelings of compassion, unity, and—dare I say it—love. If our brains are wired for religious and spiritual feelings, the monks have got it down.

But if Muir and Emerson and, before them, eighteenth-century Irish philosopher Edmund Burke had it right, feelings of spirituality don't just spring from religion: they also spring from transcendent experiences in nature. In 1757, the twenty-eight-year-old Burke landed in the center of the Enlightenment when he published *A Philosophical Enquiry into the Origin of Our Ideas of the Sublime and Beautiful*. A secularist, he'd been rambling around Ireland and feeling, for lack of a better word, moved. Sensitive and dramatic, he was less interested in landscapes that were picturesque than in scenes that were a little bit dark. Haunting was good, terrifying even better. "The passion caused by the great and sublime in nature," he wrote, "when those causes operate most powerfully, is Astonishment; and astonishment is that state of the soul, in which all its motions are suspended, with some degree of horror." He loved a torrential waterfall, a violent storm, a dark grove. He would have made a good raft guide.

According to Burke, for something to be truly awe-inspiring, it must possess "vastness of extent" as well as a degree of difficulty in our ability to make sense of it. That awe also inspires feelings of humility and a more outward perspective has been well described by philosophers, priests and poets. Until Burke, awe was considered the purview and foundational emotion of religious experience. The word "awe" derives from Old English and Norse words for the fear and dread one felt before a divine being. It isn't for nothing that many churches play up the music, the visions, the robes and architectural heights and spans. These elements fill us with wonder, humility and a bit of trepidation.

In liberating the feeling of awe from the fabric of religion, Burke heavily influenced Kant, Diderot and Wordsworth, who all wrote of the power of the sublime to shore up the imaginations and mental perceptions of humans. In America, Emerson picked up Burke's themes of vastness and humility, writing in his famous essay "Nature" in 1836, "Standing on the bare ground, my head bathed by the blithe air and uplifted into infinite space, all mean egotism vanishes. I become a transparent eyeball; I am nothing." That secular transcendence still informs the modern environmental movement.

Later, Einstein would say, "The most beautiful emotion we can experience is the mysterious." You may be rolling your eyes about now, but Emerson and Einstein were onto something. Among certain circles in psychology (those circles, admittedly, residing largely in California), awe is considered not just a powerful emotion but perhaps the sliest Power Emotion of them all. Until recently, though, there was surprisingly little scientific investigation of awe, despite the fact that it's considered one of the core positive emotions, along with joy, contentment, compassion, pride, love and amusement.

"Basically, awe is something that blows your mind," Paul Piff, a psychologist at the University of California, Irvine, told me. There are degrees of awe, he explained, from the momentary amazement of watching weird dancing-toddler videos on Facebook to seeing Northern Lights for the first time, which can reconfigure your view of the universe. A deeply powerful, awe-inspiring experience can change someone's perspective for a long time, even permanently.

Roland Griffiths is a psychopharmacologist at Johns Hopkins who studies the sometimes profound, awe-filled experiences of terminally ill patients who ingest psychedelic substances. It's not unusual for them to hallucinate they are leaving their bodies, flying over landscapes and encountering divine beings. Griffiths told journalist Michael Pollan he considers these mind-trips a kind of "inverse P.T.S.D."—"a discrete event that produces persisting positive changes in attitudes, moods, and behavior, and presumably in the brain." This is how entranced astronauts describe the "overview effect" when viewing the earth from space. Awe-triggering, life-shifting jolts are also recounted by survivors of near-death experiences and by more prosaic mountain climbers, surfers, watchers of eclipses and people who swim with dolphins, among others. When they are vast, nature scenes and events can connect us to deeper forces in the world. At the very least, these types of experiences appear to alter us temporarily.

To find out how, Piff, Dacher Keltner at UC Berkeley and two other colleagues conducted some unusual experiments. Keltner had already posited that awe is a unique emotion that turns us away from narrow self-focus and toward the interests of our collective group. To see if awe makes us more generous to each other, the researchers asked 1,500 people how much awe (and other emotions) they experienced on a regular basis. Then they gave some participants ten lottery tickets, telling them they were free to give some away to people who didn't get any. The researchers found that people who reported experiencing the most awe gave away 40 percent more tickets than those who reported the least feelings of awe. Those who experienced other emotions didn't behave more generously.

Next, they attempted to induce awe in real time by taking subjects to a tall grove of Tasmanian blue gum eucalyptus trees, and asking them to look up for one minute. They sent other subjects to look up at a tall science building. In both settings, a lab assistant "accidentally" dropped a handful of pens. Even after just one minute of awe, the tree-gazers were more helpful, picking up more pens on average than their counterparts.

But in one of the most provocative studies of all, Keltner and colleagues asked participants how many times in the previous month and on that very day they experienced up to twenty negative and positive emotions such as fear, anger, joy, surprise, etc. They also took saliva samples from the subjects and measured their levels of cytokine IL-6, a marker for inflammation. Part of the immune system, these signal molecules help heal wounds and fight illness. In healthy people, lower levels are considered better, while chronic high levels have been linked to depression, stress and poor muscle repair. Of all the positive emotions, experiencing awe was the only one that predicted significantly lower levels of IL-6. Why would this be the case? Keltner posits it's because awe causes us to reinforce social connections, which are in turn known to lower inflammation and stress. Awe wants to be shared.

Not all awe is positive. But even really scary awe—the kind that happens when a hurricane or a twister levels your town—has a remarkable ability to spur people to help each other and to unite a community toward common goals. It's evolutionarily adaptive to reach out and connect when confronted with vast forces we don't totally understand. That's how we get by.

———————

DARWIN CONSIDERED empathy or compassion to be our strongest instinct, one that launched the success of the human species. By taking good care of each other, we thrived through long childhoods, sicknesses and food shortages. Berkeley's Keltner argues we possess a literal seat of empathy: the body's vagus nerve. It starts on top of the spinal cord and tentacles out to facial muscles, the heart, lungs and digestive organs. A key switch in our parasympathetic nervous system, the vagus slows down our heart rate after a fright, bringing us back to a place of conciliation rather than aggression. It appears connected to our oxytocin receptors, which regulate the neurotransmitter that is sometimes reductively called the love hormone, since it flows during sex and breastfeeding. During the release of oxytocin, the vagus nerve may trigger an electric, humming sensation in the upper back. It's like getting electrocuted by love.

As it responds to love, posits Keltner, the vagus nerve also responds to awe. To get a better handle on how it may work, Keltner and Craig Anderson, his graduate student at Berkeley, invited me (and a whole lot of research subjects) to sit down and watch some of the most awesome video footage they could find—the earth from outer space. This is the view that caused astronauts to fill with tenderness for their little marble in the sky and all of humanity on it. This sensation may be close to what Buddhists describe as the unity of Nirvana, a transcendent happiness characterized by outward love and the elimination of desire.

Unfortunately, I wasn't approaching Nirvana while watching the earth-from-space footage on a smallish video monitor in a utilitarian lab at Tolman Hall. Anderson had me strap on a heart-rate monitor, and then he attached sensors to my finger for measuring skin conductance (sweating, another measure of the autonomic nervous system). He launched that video, followed by one of magnificent mountain summits, and after ten minutes or so, he returned with my results. In accordance with his overall study data, my heart rate did decline while viewing the monitor. But not much happened with my skin conductance, nor with my facial muscles that Anderson had been surreptitiously monitoring with a hidden camera.

As to why the heart rate slows while viewing the sublime, Anderson has a theory. "Things that cause people to feel awe tend to be information-rich, vast, and things that we have trouble wrapping our minds around," he said. "So basically, the body is quieting down a bit so that it can take in information in the environment."

My vagus nerve did not seem to get the memo. I was not even experiencing one of the telltale signs of awe that is surely one of the best words in science: piloerection, or hair standing on end. Sitting in a cubicle with electrodes sticking out of my finger, I did not feel like I was hurtling through deep space, nor, as Strayer had argued in Utah, is watching videos of nature much like the real experience of standing in some enormous viewshed taking in the sensory gifts of the biosphere. In fact, perhaps the absence of scale-induced awe is one of the reasons virtual nature will likely never match the real thing. Burke's essential ingredient of vastness is hard to simulate on a screen, although a background soundtrack by John Williams certainly helps.

Among other things, awe promotes curiosity, explains Anderson. This is because we experience things out of our normal frame of reference, things we can't easily categorize or understand. When we are curious, we are drawn out of ourselves. We seek information from others. With their mixture of fear, beauty and mystery, these experiences also tend to get seared into our memory. I will probably never forget seeing my son's face for the first time, or peering into the Grand Canyon as a child, or watching Northern Lights swirl in an Alaskan sky or driving through a surreal lightning storm in Texas.

We can also experience awe before very charismatic individuals like cult leaders, celebrities, kings and fascist dictators, who embody a vastness of skill or might and are wise to cloak themselves with the trappings of status and an air of inaccessibility. Awe channels power. Melanie Rudd, who studies consumer psychology at the University of Houston, wanted to know if awe, by focusing our attention on the present moment, might expand our perception of time. Anything that could do this might be a great discovery "given that there is a huge time famine in many societies in the world," as she put it, "and this has a huge impact on mental and physical health, life satisfaction, depression, headaches and hypertension." Nearly half of all Americans feel they do not have enough time on a daily basis.

When Rudd induced either awe or happiness in her lab subjects, only awe led them to feel less time-pressured, to report less impatience and to volunteer extra time to help others. These happened after quick interventions, such as looking at videos of whales and waterfalls, suggesting that images can indeed induce at least some feelings of awe. The implications of her work are huge for consumer advertising. Seen an ad lately for a new car? Chances are it's traveling through a magnificent landscape, not stuck on the Beltway. "Lots of things we buy can get framed in an experiential way," she said. "Being in nature had the biggest effects we saw."

Very few studies have looked at awe and behavior in the field, other than Piff's one minute of staring at trees. But if we look at our phones (don't tell Strayer), it's evident that people want to share experiences of awe. That's why we Instagram photos of sunsets and "like" videos of swarming starlings while savoring another great word: murmuration. We now experience small moments of awesomeness on a daily basis through our feeds and our screensavers. Perhaps these "microbreaks" help make up for the loss of the powerful and the vast connections to nature we used to experience when we spent more time outside, but "the jury is still out on how much social media shapes our everyday experience of well-being," said Irvine's Piff.

The fact that a discussion on awe finds itself circling back—like so many discussions—to our technology, made my three days unwired in the desert feel all the more radical. We've got awe! We've got it live right here in the ancient handprints and the umpteen gazillion stars and the fact that a nerdy bunch of students will head back to the city with new friends and a new way of looking at past and present.

As to whether any of this will show up in our cranial currents, initial results seem promising. Strayer sent me the results from my wired-up river interlude, and they were consistent with his hypothesis. A colorful graph showed the power of my theta waves at a range of frequencies compared to samples from the two groups that stayed in the city. My theta signals were lower, indicating a prefrontal cortex on a brief vacation. What the graph doesn't tell us, though, is exactly where that energy is going in the rest of the brain. Although Strayer-the-Scientist wants to keep unpacking the signals like a Matrushka doll, Strayer-the-Mountain Man understands some mystery will remain, and that's okay.

For millennia, humans alone or in small groups have at times sought out a sparer, more elemental connection to the forces of nature. They come because they are needing something, and they keep coming because they are finding it. Their pursuits may be spiritual, interpersonal or emotional, deeply human and complex and unlikely to be explained in a bar graph. "At the end of the day," said Strayer, his eyes grazing the horizon, "we come out in nature not because the science says it does something to us, but because of how it makes us feel."

Water on the Brain

"Oh, Eeyore, you are wet!" said Piglet, feeling him.
Eeyore shook himself, and asked somebody to explain
to Piglet what happened when you had been inside a river
for quite a long time.

—A. A. MILNE

Between every two pine trees there is a door leading to
a new way of life.

—JOHN MUIR

The first American veteran on Idaho's main stem of the Salmon River was Captain William Clark. He and Lewis had split up to search for a route to the Pacific, but this one was not panning out. The rapids were too rough for the expedition's 1,000-pound dugout canoes, and the canyons proved too steep to portage around the whitewater. After exploring the upper reaches in moccasins, Clark complained, "I Sliped & bruised my leg very much on a rock." With that, he etched his name in a pine tree and got the heck out of there. That was in 1805.

Eventually, other explorers, fortune-seekers and recluses followed, braving the rough terrain to stake mining claims downstream. True to its cognomen as the River of No Return, the river accommodated one-way traffic only. The miners built huge wooden boats laden with supplies and ventured down the rapids. If man and boat survived the passage, the boat would get cannibalized for a cabin and the miner would lay in for a long, long while.

The steep country that hemmed in the river was never ideal for human habitation. In 1980, Congress made the isolation official, designating the river and its surrounding mountains the largest chunk of the wilderness system in the Lower 48. The Frank Church–River of No Return Wilderness, sometimes just called "the Frank," stretches across 2.3 million acres in the part of Idaho that starts to get skinny. The river running through it carves a long, forested gorge deeper than the Grand Canyon.

It was through that gorge that another group of American veterans—all women, all scarred emotionally and physically by their service—descended in the summer of 2014. Like Clark, they were also on a voyage of discovery in the American wilds. I wanted to witness it. If one minute of gazing up at a eucalyptus tree makes people more generous, and three days makes them more socially connected, calm and inspired, what could a week unleash? Were the inverse-PTSD effects of awe real and if so, would they show up in the brains that needed them most?

YOU HAVE TO be brave to venture down the Salmon, and a little bit addled. This group of women, sponsored by an Idaho-based non-profit called Higher Ground, was both. Participants had to be former or current members of the military who suffer from PTSD, posttraumatic stress disorder. When I learned the organization was willing to invite a journalist, I signed on.

This was Higher Ground's first all-women's river trip. The plan was to float eighty-one miles of the river, try our skills at kayaking, rowing, and paddleboarding (nonmandatory), participate in "processing" groups and team-building activities (mandatory), eat together, collapse into tents, and then do it all over again the next day. On the sixth day, we'd leave the river, flying home off a dirt strip in small planes. Unlike the miners, we'd be returning to civilization, hopefully a little bit changed.

The night before launching our boats at the end of a dirt road, I met up with the women, gathered on a restaurant patio for pizza in the no-traffic-light town of Stanley, rimmed by the vaulting, aptly named Sawtooth Mountains. This clearly was not your usual river-rat crowd. These women were on the whole younger, more ethnically diverse and less able-bodied. The nine former service members carried an assortment of cigarettes, butch hairstyles, tattoos, piercings and physical supports that included a cane, orthopedic tape and an arm splint. Collectively, they brought a small pharmacy's worth of antianxiety drugs, antidepressants, antiseizure meds, painkillers, digestive aids and sleeping pills. One service dog, Major, a yellow lab mix, wore a bib that read DO NOT PET. The warning could have applied to anyone. Heavy-lidded and surly after a long day of travel, they were not about to smile for a bunch of cowtown selfies.

The recreation therapists, Brenna Partridge and Kirstin Webster, handed out matching black fleece jackets emblazoned with the unique crest of this "unit"—HG-714-RA, which stood for Higher Ground, July 14, Rafting. (Other Higher Ground units, typically coed or all-male, might spend a week fly fishing or skiing or doing lake sports.)

Partridge smiled and asked us to introduce ourselves and talk about why we wanted to be on the trip. Marsha Anderson (some names, including hers, have been changed) described being medevaced out of Afghanistan on a stretcher, convinced for a while that she was already dead. It took her thirteen months to relearn to walk. Now she felt angry, misunderstood by her family, and cheated of the sports she loved like surfing and cycling. She was hoping to find some new ones, along with new friends who had been through what she'd been through.

Carla Garcia, thirty-five, described how she'd volunteered for the first Iraq invasion in 2003 and then returned as a vehicle commander running fuel convoys across the war zone from Al Taqaddum. In 2005, her truck hit a roadside bomb and she was blasted from it, landing on her head. Her driver died. During her third tour, in Mosul, another bomb exploded, crashing her head against the vehicle roof and pelting her with shrapnel. Garcia pulled her ailing driver from the smoking wreckage and fought off insurgents with her M-16 until she passed out (she received both a combat action badge and a Purple Heart, I found out later). Doctors induced a week-long coma to relieve pressure in her brain. Afterward, she had to learn how to talk. In addition to chronic pain, she suffers seizures, headaches, mood swings, and nightmares. She can't walk far, won't drive, and can barely stand being in any kind of vehicle. "I don't like crowds and I don't like people," she said. "This will be hard."

After dinner, we grouped for the processing talk, our first one, to articulate goals for the trip. That's when Kate Day, a Navy vet in her fifties from Las Vegas, mentioned her three-year stint of homelessness, a stay in a mental institution and her near-inability to leave her house. Two other women chimed in that they too had been institutionalized. One said she was still so depressed she didn't want to keep living. Another said her anger and misery had alienated her whole family. Another, sitting expressionless, said in a flat voice that she wanted some time to "be in the moment and not zone out." A skinny blonde wearing a sparkly blue sundress and pink sunglasses, whom I'll call Pam Hana, showed the opposite affect: manically chatty, never still. She woke up scared and crying because she hated airplanes and had successfully avoided them for years until this trip.

Tania Herrera, wearing a Gilliganesque fishing hat under dark, cropped hair, talked about being limited by her body. First struck by shrapnel near Fallujah, then catapulted by a car bomb along her convoy route and finally struck by pieces of a collapsing mosque hit by a grenade, the former Army gunner now had one working arm, a bad leg and a brain that didn't work too fast. Thirty-four years old, she rarely left her house near Fort Bragg. "It sucks to think that's the way life is going to be, stuck in a rut," she said. "It seems like a life sentence."

Petite with smooth skin, and a friendly, wide mouth, Herrera also told us that she now had trouble making friends, and on top of that, she had some serious hair issues. "I used to have long hair but can't figure out how to do it with one arm," she said. "I used to sit on my hair like Medea. I'm not that girly, but to have it stripped away from you is hard. I don't want to go to family weddings because I can't look pretty."

Partridge, the group leader, gave Herrera her marching orders: "Find someone to bond with. This is your unit now."

In the days following, more details of their battered lives would come out during processing, in one-on-one talks or in small groups. As a general rule, the younger women had seen combat, even though technically they weren't supposed to be in combat roles at the time. That was a central irony of serving in recent wars, and yet, because they were women, it was often harder for them than for men to get diagnosed as having combat-related PTSD. Many of the older women were here because they suffered military sexual trauma (MST). One was gang-raped by eight men, including her commanding officers, while stationed in Okinawa; another was attacked in the Navy by her master-at-arms. Another was assaulted by a civilian while on leave in Europe. In only one instance did the perpetrators meet justice, and that was the civilian.

In both types of PTSD, the consequences are similar: life-altering social, professional and psychological impairments.

EVERY BIG WAR has its signature wounds. If the Civil War didn't kill you, you were likely to end up with amputations. Surgeons in World War I advanced the art of facial plastic surgery (mustard gas liquified facial tissue). Gulf War veterans barely saw combat, but many suffer from mysterious symptoms believed to be linked to nerve agents. PTSD was common after most of these wars— even Homer wrote about it—but it went by different names: shell shock, soldier's heart, combat fatigue. Frederick Law Olmsted, from whom I have a quote in just about every chapter (because, as well as being a badass nature guru, he was, like Zelig, witness to just about every significant beat of the nineteenth century, from plantation slavery to the gold rush to the invention of suburbia), described the Union soldiers after the Battle of Manassas as a "disintegrated herd. . . . They start and turn pale at the breaking of a stick or the crack of a percussion cap— . . . It is a terrific disease." PTSD wasn't officially named and recognized by the Veterans Administration until 1980.

In the general population, about 8 percent of us will experience PTSD. Among veterans, that figure is about 18 percent, but a recent examination of the data for over a million veterans of the wars in Afghanistan and Iraq found a 27 percent rate (with over 70 percent of that coexisting with depression). The fingerprints of the recent wars are so far clear: PTSD, traumatic brain injury (TBI) from explosives, and sexual assault.

Some studies suggest that women experience PTSD at slightly higher rates than men, or they may just more readily admit to having it. According to the latest iteration of the *Diagnostic and Statistical Manual of Mental Disorders*, symptoms of PTSD cluster around four subgroups: reexperiencing (flashbacks, nightmares), avoidance and withdrawal, bad moods and depression, and hyperarousal, such as jumpiness, vigilance, aggression and sleep problems. Women, who now make up about 15 percent of the military, express some symptoms differently, experiencing higher rates of anxiety and eating disorders. They are two to four times more likely than other women to be homeless; men have more problems with violent aggression and substance abuse, but plenty of women experience these also.

By all indications, the women on our trip were like Tania Herrera, who'd been an eager, straight-A high school student from North Philly: highly competent, enthusiastic recruits when they started out. Their intelligence and toughness were still in evidence. But pieces of them had crumpled. They no longer felt whole, or secure, or capable. Now they were grieving their lost selves. As Herrera put it during one group session, "I never thought I'd be thirty-four and unable to take care of myself. When I went to war, I thought either you die or you make it out. I didn't factor in what if you came out different than when you went in."

The women described daily lives involving constant physical pain. They couldn't concentrate well. They were sometimes jumpy, depressed. They didn't like being with people, but they didn't like being alone all the time. The wars had taken away their ability to sleep well.

IT WAS TIME for the women to get out of their lives and into the river. The first named rapid, called Killum, came up fast. I was paddling one of the four inflatable kayaks, and I saw the kayak in front of me meet a short wall of water and flip. I hit the same cold sideways wave, my paddle dove, and I flipped too. Happily, the Class II and III rapids in this stretch are more wave than rock, and they are short, interspersed with deep, calm stretches. I managed to claw back into my boat. The six women paddling the raft cheered me and the other kayakers on.

Many rapids followed before camp, and I was alternately exhilarated, nervous, cold and determined in that I'm-committed-now kind of way. Entering a rapid, your vision narrows and so does your focus. Your heart rate picks up, your breath quickens, and your skin temperature rises. Your gut begins to tighten. In small doses like this, the adrenaline rush is fun. You feel present; the B roll of your mind falls away, and there's a heady release of endorphins when you're safely through. Kayakers sometimes call paddling big water "combat boating," and when hard-shell boaters roll their overturned kayaks back upright while still strapped in, it's called a combat roll.

I saw the basic inanity of this metaphor while surrounded by these very real veterans. In war combat, the stress response isn't small or ephemeral. It's big. And it lasts for days, sometimes weeks or months. It lasts so long that the brain changes—more in some people than in others. Blame evolution. Our nervous systems are naturally hardwired for fear, telling us what to avoid and how to stay safe. Some psychologists argue that fear is our oldest emotion, existing in the earliest planetary life forms and predating even the drive to reproduce. It starts deep in our brainstem, in the Milk Dud–sized amygdala.

When fear alone rules us, we lack the smarts to do much of anything creative, or interpersonal, or spatially demanding. Part of what makes us human is that our brains evolved a neocortex, the place where we plan and puzzle and tell ourselves we're being drama queens. A fright causes a neurological tug of war between the old and new brains. In the deep clutch of fear, our primitive brainstem overrides our problem-solving neocortex, and we become stupid. With PTSD, the brain stays locked into amygdala hyperdrive. Failing to bounce back to baseline, it loses the ability to distinguish between a real and a perceived threat. That's why soldiers with PTSD often cannot tolerate driving or shopping or loud noises even in safe places when they return home.

But there's a reason we feel fear. It may have given us the gift of memory. The very reason we remember anything may be that we must remember near-misses, narrowly avoided dangers, and attacks from predators and enemies. Thanks to fear, we enjoy the smell of madeleines and the writers who write about them.

At its root, PTSD is a memory disorder. Brain scans of people suffering PTSD show cellular and volume changes in the hippocampus, a region that helps process memories and sits very close to the amygdala. In frightened lab animals, the fear hormones—glucocorticoids like cortisol, norepinephrine or adrenaline—flood receptors on the hippocampus and impair memory. It appears that persistent trauma memories shrink the hippocampus, and it's well established that PTSD leads to emotional as well as cognitive problems, such as poor focus and short-term memory deficits.

Physiologically, chronic, heightened stress looks like this: higher blood pressure, cellular inflammation, and a higher risk for cardiac disease. Longitudinal studies show that veterans with PTSD are sicker, in more pain, and die younger than their non-PTSD peers. They are also 4.5 times more likely to have substance abuse issues. Veterans are twice as likely to be divorced, and female veterans commit suicide at nearly six times the rate of other women.

Groups like Higher Ground—and there are many, from those offering surfing and fly-fishing programs for vets to a hospital in Los Angeles that promotes bonding between humans and abused parrots with symptoms resembling PTSD—believe that engaging with nature or wildlife can reduce trauma symptoms. Adventure sports like kayaking provide a laser focus for an unfocused mind, as well as a welcome distraction from unwelcome thoughts. The physical exertion often leads to better sleep, and, as we've seen in previous chapters, the sensory elements of nature can calm the nervous system.

Even knowing all this, I couldn't help but worry a bit about these women in such an uncontrolled environment. What if they got pinned on a rock or had a bad swim? One of the kayakers was Marsha Anderson, who'd been a ski racer in her youth in Wyoming. Now she had nerve damage in an arm and a leg from an explosion in Afghanistan in 2009, and she hurt all the time. After her injury, she couldn't walk for a year. She seemed fragile. When a rapid spat

Anderson out of her inflatable boat in midafternoon, I held her boat next to mine and helped haul her back in. Then Herrera, riding in a double kayak, her right arm bearing her high-tech brace with a GoPro camera attached to it, went over. I wondered how she would get back in the high, slippery boat with one working arm; but her partner, a Higher Ground staffer, stayed in the river and heaved her over the gunnels.

If these women came expecting a relaxing repose on the beach-lined river, this wasn't it. We weren't even allowed cocktails. Could they handle this sort of extreme adventure? These women lived in a constant playback of memories and anxieties. Maybe they should be home snuggling with their service dogs and using a rowing machine?

Or maybe not. Anderson, a Korean American in her early thirties with short hair, sat smiling while she ate an eggroll that evening. "I never thought I'd go by myself down a river," she said. "I'm exhausted from the adrenaline." She recalled the words of a yoga instructor: "Anxiety is just excitement without breath." The river was teaching her to breathe. "I wasn't sure I was going to go back in and keep kaya-king," she continued, "but I did, and I was trying to breathe in every rapid." She clearly liked being a badass. Who doesn't?

As for Herrera, who was still relearning how to take basic care of herself, paddling a kayak was a revelation. She didn't seem to mind the unplanned swim at all. She found that she could tape her bad hand around the paddle shaft and use the other arm for most of the power. Seeing her in the boat, I was reminded of another one-armed veteran who made a similar river voyage 145 years ago, Major John Wesley Powell. Wounded during the Civil War and commissioned to survey the frothy Colorado, he seemed to relish every minute of it: "We have an unknown distance yet to run; an unknown river yet to explore. What falls there are, we know not; what rocks beset the channel, we known not; what walls rise over the river, we know not."

When Herrera flipped, she even had the presence of mind to salvage her combat-medal-bedecked Gilligan hat. "I was really happy that I was able to contribute and not have everyone do work for me," she said. "It was neat to do something physical. At home, I can barely get my own mail."

The rafters, too, had a good day. Anjah Mason, the expressionless Army vet who had told us she wanted to stop zoning out, described almost having a panic attack on the boat, but then talking herself through it. She'd learned how to adapt to a wholly new situation, and she was pleased.

Everyone was hungry. No one stayed up late. Manic Pam Hana finished a cigarette and then fell asleep in front of her tent under the still-bright northern Rockies sky at 8 P.M.

I began the next day with my signature outdoor ailment, a bee sting. Catalina Lopez administered rubbing alcohol and Benadryl and told me to keep tabs on the swelling. A former Army nurse, she had served for fifteen years in the Balkans, Somalia and Iraq, and was haunted by recurring dreams of blood and severed body parts. Once, while I was eating lunch, she had described watching an unconscious guard's brain swell and swell in the hospital. She told me normal intracranial pressure was 10, but this guy's meter was reading 20, then 30 and then 85 "and then I could see his cranium start to move."

I looked at my sandwich.

"You see where I'm going with this."

I nodded.

"Do you want me to stop?"

"Yes, please."

That second day, I joined the increasingly sociable paddle raft. At some point, Tania Herrera, sitting on the raft tube across from me in the bow, started singing, "I kissed a bug and I liked it." She told stories about being in Iraq. She was part of an all-woman transport truck they nicknamed the Maxi Pad. Then someone asked me why I wanted to write about breasts, the topic of my first book. That inspired Herrera to come up with a name for our rubber boat: the Boob Tube.

It was a long day on the river, a hot, twenty-mile paddle punctuated by swimming and a beachside lunch. The canyon in this section is steep and dotted with large ponderosa pines that emerge from shiny black gneiss outcrops. We were passing through the middle of the ancient Idaho batholith. Angela Day, a blond, plump Navy veteran, bobbed along in her kayak like a mellow duck, not working too hard and giggling through the waves. Anderson, the nerve-damaged former ski racer, rode the stand-up paddleboard; in the rapids it became more of a kneeling board, and sometimes an upside-down board. In the afternoon, nurse Lopez spilled out of her kayak in a tricky rapid. From the Boob Tube, I could see the panic in her face, the desperate gulps of air and water. She got back in the kayak, but she wasn't happy about it.

At processing that evening, she looked defeated. Facilitator Partridge had asked the group what their passions were. "I used to be passionate about everything," said Lopez, whose PTSD and a chronic back injury got her medical retirement from the Army. "Life, work, nature. Even today I was passionate about kayaking until, what the fuck, and now I expect to be disappointed by everything." She shrugged. "Maybe I'll get back in, I don't know."

Anjah Mason said she didn't know what she was passionate about. "I used to be passionate about my family."

Connie Smith, a former Navy captain from Texas, said she was passionate about her work training service dogs.

Angela Day said she was passionate about her relationship with the Lord. "Today, in the kayak, I was like, 'Come on, Lord, bring it on! You can do better than that!'"

Linda Brown, soft-spoken, in her fifties, said she was passionate about outdoor sports. "I can't say I'm passionate for any length of time, but I do believe I'm passionate about the outdoors, trees especially."

Pam Hana, still manic, bouncing on her chair, said, "I'm passionate about staying single and frickin' free! I'm loving it! Seriously!"

Herrera said she used to be passionate about her job in the Army. "I was a lead gunner in Iraq, in a turret, with a headset. My kid dream came true, of a car that talked to me. I wanted to be Knight Rider with the biggest gun and the coolest clothes. I remember thanking God for allowing my dream to come true." She looked at the sand. "It's so hard to create a dream again and go forward. That's where I get stuck. How do I do that now with all these barriers, these health issues, the medicines, the bad relationships, no money, the disability?"

Angela Day said, "I don't want to leave my safety zone."

"You left it today on the river," Partridge said.

"Yes. But it's become normal for me to leave the house only once a month to buy groceries. I do have a deep personal dream not to be that way."

"Like when you're on the river, sometimes you have to ask for help," said Partridge. "People have your back."

"It was the funnest day ever today!" said Hana.

"For you." Lopez glowered.

WE FELL INTO a pattern of running the rapids, processing the day, making and breaking camp, telling stories, coming together sometimes and other times dropping off into pockets of introspection or quiet or just plain tiredness, not unlike riffles and eddies, the rhythm of the river. Before breakfast, we practiced group yoga. Several people grabbed a quick cigarette before they arranged themselves into lotus position, which never failed to crack me up. Day's dog, Major, lay at her feet at all times and seemed perplexed by the strange body positions. Even impassive Mason, who moved as little as possible, swiveled her torso for gentle twists. Skinny Hana was usually cold but always smiled. I noticed she was babbling less.

Each day, there was more laughter. Lopez created nicknames for our guides, who rowed the gear boats, cooked the food, set up our tents and then left us alone. They were all young, strong, and mostly male. She dubbed the clean-cut trip leader, Reid, Captain America. Another, burlier guy with long hair became Fabio. Like an army unit but with better hair, they had their jobs, routines and ways of contributing to the group. Some were funny, some wise, some watchful.

"This is not unlike war," Herrera told me. "There's something that can kill you. There's a tight group that depends on you for survival, and everyone is a part of it. Bonds develop that have meaning. Life is better when it's simple. Here, like in the Army, you don't have forty different options for toothpaste. You have your place. We all have it."

IT'S NO WONDER there is a storied American legacy of damaged soldiers heading for the wilderness. The backwoods of Idaho, Montana and Alaska are notoriously peopled by veterans. After Vietnam, men went there who felt misunderstood by civilization and found the greatest peace away from it. But despite its strong anecdotal legacy, the wilderness is not recognized by the Veterans Administration, or even by most psychologists, as a legitimate healing tool. It's largely been the veterans themselves, privately funded and socially inclined, who are driving the current renaissance of programs aimed at helping service members.

David Scheinfeld, who has led Outward Bound backpacking courses for veterans for 11 years, uses the term "therapeutic adventure," but doesn't necessarily share that with the participants. He has seen so many lives transformed by six-day trips in the wild that he decided to study them for his Ph.D. in psychology from the

University of Texas, Austin. He wanted to know what was making the approach successful while the other standard interventions— cognitive behavioral therapy and medication—were falling short.

When Scheinfeld assessed 159 veterans, he found that Outward Bound participants experienced 9 to 19 percent improvement in mental health; vets in the control group showed no such gains. The Outward Bound groups still showed the boost a month after the trips ended.

Why did the trips work? Scheinfeld noted that the participants, mostly male, tended to encourage each other to give counseling another try and stick with it. "There were always a couple in each group who were helped by counseling, and they became de facto mentors," he said. Because of this, the participants showed a greater openness seeking treatment after the trips, and they were less likely to drop out of treatment. Another reason for success: the trip itself, being in the wilderness and part of a supportive group, counted as extended therapy, and for many hours a day, not the usual one hour per week offered by the VA. "It's hard for these guys to sit in a room with four walls and talk about their feelings," said Scheinfeld. "It just happens best when they're in a natural setting. It draws them out."

Other studies have shown similar results. Neil Lundberg of Brigham Young University looked at twenty-two participants from two Higher Ground trips in 2010. Compared to a similar group of veterans on a waiting list, they showed significant decreases—of up to 40 percent—in flashbacks, emotional numbing and hyperarousal after the trips. But not everyone is convinced. Craig Bryan, a psychologist and Air Force veteran who directs the University of Utah's National Center for Veterans Studies, remains skeptical of nature-based treatment. Most of the studies out there, he said, are small, lack a meaningful control group, and don't follow participants for very long. "It's possible these treatments are better than existing treatments, but we just don't know," he said. "We don't have the data to back it up. I want to see randomized control studies, bigger studies."

To collect more data on its programs, Outward Bound is partnering with the Sierra Club and the U.S. Department of Veterans Affairs to run a large pilot study out of the Seattle Veterans Administration. Between them, Outward Bound and the Sierra Club reach hundreds of veterans per year. Stacy Bare is helping to coordinate the study for Outward Bound. A veteran who credits his time in the wilderness with saving his life, Bare understands the need for better metrics.

"It's amazing to me that we don't know more," said Bare. "I think we all believe in the power and mystery of the great outdoors, but these are difficult things to quantify by science. Is it difficult to do a double-blind control study in nature? Very. I don't think we have to hit that standard, but we have to have a more systematic approach to how we evaluate the effects of the outdoors."

DURING THE LAST days on the river, we floated through a landscape that had been ravaged by wildfire in 2000 and again in 2012. At the site of the older blaze, new teenaged evergreens were rising. Around the charred stalks of the more recent fire lay a carpet of brilliant green grass. It was a powerful reminder that life cycles onward. One morning I sat on a big gear raft next to Linda Brown, the older vet who had been institutionalized for depression. She sat with her arms wrapped around her life jacket, her sandaled feet propped on the front tube of the boat. "The trees can't control their lives," she'd said, speaking so softly she practically whispered. "We can't always control what happens to us. The trees can teach us acceptance. And metamorphosis."

Months later, most of the women of Unit HG-714-RA would look back and say rafting in Idaho helped them on their long journeys to recovery. At least one of them, Catalina Lopez, nurse of exploding heads, would say it didn't. Statistically, this seems about right. In other mental-health studies, for example in Finland, about 15 percent of subjects remain wholly unmoved by their time in nature. Sometimes it's because they just hate it there. They hate the bugs, the breeze, the big sky. Their nervous systems will never calm down outdoors.

That wasn't Lopez's problem. She said the trip just wasn't nearly long enough. Not long enough for her to turn off her nightmares. Not long enough to stop her from sleep-driving through corn stubble at midnight on Ambien. Not long enough for her to start believing again in other people. Certainly not long enough to gain confidence swimming through swift currents. Many wilderness therapy programs for troubled adolescents run weeks and weeks, even months.

Although Higher Ground gives each participant a "recreation fund" to keep pursuing an outdoor sport at home, Lopez told me she still hadn't decided whether she would use it. But Marsha Anderson and Carla Garcia would go surfing, sometimes together, in Southern California. Formerly passive Anjah Mason had joined a gym, determined to lose twenty pounds. I was amazed at her transformation. She now routinely hikes near her home, and she wanted camping gear. Pam Hana had been cycling and wanted to use her rec funds to buy a mountain bike.

As for Herrera, she told me she was signing up for another river trip, this time with Outward Bound. "I liked the river. I liked to be successful," she said. And she was scoping out other programs as well, a shooting trip in the countryside in Alabama, maybe sky diving or rock climbing if she could find a place that adapts to disabilities. "I want to find something to do every summer," she said.

And, she told me with pride, she was growing her hair out.

Please Pass the Hacksaw

Childhood is, or has been, or ought to be, the great original
adventure, a tale of privation, courage, constant vigilance,
danger, and sometimes Calamity.

—MICHAEL CHABON

By second grade, it was clear that although Zack Smith could sit in a chair, he had no intention of staying in it. He was disruptive in class, spoke in a loud voice and had a hard time taking turns. His parents fed him a series of medications for ADHD, attention-deficit hyperactivity disorder, many of which didn't work. Zack, who attended school in West Hartford, Connecticut, was placed in special classrooms, where he showed a propensity for lashing out. Twice suspended, he was miserable. He didn't seem to care about anything at school. When his parents realized that his path would likely lead to worse trouble, they pulled the ripcord on eighth grade.

Where Zack eventually landed was spread-eagled on an east-facing slab of quartzite in Pendleton County, West Virginia. His chin-length, strawberry-blond hair curled out beneath a Minion-yellow helmet. A harness cinched his tee shirt—the sleeves of which have been ripped off—obscuring the *Call of Duty: Advanced Warfare* lettering.

"I have a wedgie!" he bellowed out from 20 feet up.

Belaying him was another fourteen-year-old, scrawny, earnest Daniel. Earlier in the day, Daniel had asked, "Do I have to belay? I'm only ninety-five pounds." Both kids still looked a little apprehensive, but there was no question they were paying full attention to the wall of rock and to the rope that united them. Yesterday in "ground school" under a picnic awning in a campground near Seneca Rocks, they and twelve other scrappy kids from the Academy at SOAR learned how to tie figure eights and prusiks, the knots that would save their lives. Their ages spanned five years, but crossing this vast spectrum of puberty, the younger kids looked like they could be the square roots of the biggest ones. Physically, Zack occupied an awkward middle ground, lanky and knock-kneed, sporting an alarmingly deep voice behind a crooked smile.

He gradually moved his right foot to a new nub and pulled himself higher. He scrabbled upward, finally victoriously slapping a carabiner on the top rope before rappelling down. "Oh man, my arms hurt," he said at the bottom, his pale cheeks flushed from sun and exertion. Daniel accidentally stepped on the climbing rope and, per the rules, had to kiss it. This happened so often no one remarked on it. For a moment both boys cheered on Tim, a small boy from Atlanta with eyeglasses so thick they looked like safety gear. The aspirational name tape on the back of his helmet read T BONE SIZ-ZLER. A group chant began: "Go Tim go-oh, go Tim!"

Before enrolling in the outdoor adventure-based boarding school for grades seven through twelve, Zack, like a lot of these boys, had already spent some summers at SOAR, a well-established camp based in Balsam, North Carolina, for kids with ADHD and related learning disabilities. Its founding principle—radical several decades ago and still surprisingly underappreciated—was that kids with ADHD thrive in the outdoors. Since then, ADHD diagnoses have exploded—to the point where 11 percent of American teens are said to have it—while recess, physical education, and kids' access to nature have miserably shriveled.

Zack's first SOAR summer involved a three-week stint of horse-packing in Wyoming. Before the trip, he would have preferred to stay home and play video games. "I hated nature," as he put it. But something clicked under the wide Wyoming skies. He found he was able to focus on tasks; he was making friends and feeling less terrible about himself. Zack turned his restlessness into a craving for adventure—which is perhaps what it was meant to be all along.

THE HUMAN BRAIN evolved outside, in a world filled with interesting things, but not an overwhelming number of interesting things. Everything in a kid's world was nameable: foods, creatures, constellations. We were supposed to notice passing distractions; if we didn't, we could get eaten. But we also needed a certain amount of stick-to-itiveness so we could build tools, stalk game, raise babies, and plan big. Evolution favored early humans who could both stay on task and switch tasks when needed, and our prefrontal cortex evolved to let us master the ability. In fact, as David Strayer and his marching band of neuroscientists in Moab made clear, our nimbleness in allocating our attention may be one of humanity's greatest skills.

Most of our ancestors had brains that craved novelty and that wanted to explore, to a degree. This worked out for us. Our species expanded into more habitats than any creature the earth had ever seen, to the point where humans plus our pets and livestock now account for 98 percent of the planet's terrestrial vertebrates. But evolution also favored some variability in our brains, and some of us pushed exploration more than others, or were simply more comfortable in the new, unfamiliar habitats. These are the sensation-seekers among us, the ones who thrive in dynamic environments and can respond quickly to new information.

We have come to see the restlessness that was once adaptive as a pathology. A recent advertisement for an ADHD drug listed the "symptoms" to watch for: "May climb or run excessively, have trouble staying seated."

It's worth taking a look into the brains of kids like Zack, because not only do these kids need nature-based exploration, but exploration needs them. Zack and his tethered gang of misfits hold clues to the adventure impulses lurking in all of us, impulses that are increasingly at risk in a world moving indoors, onto screens and away from nature. Attentional mutants everywhere have saved the human species and they may yet spare us from Michael Chabon's dreary pronouncement that "the wilderness of Childhood is gone; the days of adventure are past." But first, we have to understand the connections between learning and exploration, childhood, play and the natural world.

If spending time in nature could be so helpful to adults, I wondered what it could mean for adolescents whose brains were still so pliant. Since kids learn everything faster than we do, it made sense that the outdoors could provide huge payoffs to kids who needed a mental recharge or a new framework for learning. Could being outside help them change patterns of emotion and attention?

The fact is, all human children learn by exploration. So I had to wonder if we are cutting them off at the knees, not just with medication, but through overstructured, overmanaged classrooms and sports teams, less freedom to roam and ever-more-dazzling indoor seductions. Modern life has made all of us, along with our kids, distractible and overwhelmed. As McGill neuroscientist Daniel Levitin explains, we consume 74 gigabytes of data every day. After school, teens now spend vastly more waking hours on screens than off them.

"The digital age is profoundly narrowing our horizons and our creativity, not to mention our bodies and physiological capabilities," said adventure photographer James Balog, even as his hard-won chronicles of a changing planet are delivered to millions digitally. Yet Balog, who roamed the hills until dark as a kid in rural New Jersey, can hardly get his eighth-grade daughter off her phone. "These are hours not being spent outside," he said. "It kills me."

It's one thing to let kids unplug and run loose in the woods in summer, but taking the whole academic year outside—the SOAR students alternate two weeks on a forested campus and two weeks in the field—reflects either parental desperation, intrepid educational insight, or a combination of the two. Zack's backstory as an institutional rapscallion is a common one, especially among boys, who are diagnosed with ADHD at more than twice the rate of girls. History is full of examples of the fortunate ones who went on to become celebrated iconoclasts like wilderness advocate John Muir, who spent his early childhood sneaking out at night, dangling from the windowsill by his fingertips, and scaling treacherous seaside cliffs in Dunbar, Scotland. Frederick Law Olmsted hated school. His indulgent headmaster used to let him roam the countryside instead. Mark Twain left school at twelve, yet clearly believed in the value of a good float trip. Ansel Adams's parents plucked their restless boy out of school, gave him a box Brownie camera, and took him on a grand tour of Yosemite. It was unschooling, California-style.

Olmsted, looking back on his life, identified the problem as the stifling classroom, not troublesome boys. "A boy," he wrote, "who would not in any & under all ordinary circumstances, rather take a walk of ten to twelve miles sometime in the course of every day than sit quietly about a house all day, must be suffering from disease or a defective education."

The Academy at SOAR—accredited for just the last three years—was determined to find a better way. The school enrolls just 32 students, 26 of them boys, divided into four mixed-age houses. Each kid has an individualized curriculum, and the student-teacher ratio is five to one. Tuition is a steep $49,500 per year, on a par with other boarding schools, although you won't find a Hogwartsian dining hall or stacks of leather-bound books. The school still covers the required academics, as well as basic life skills like cooking, but finds that the kids pay more attention to a history lesson while standing in the middle of a battlefield or a geology lecture while camping on the Ordovician formation.

"We started from scratch," said SOAR's executive director John Willson, who began working there as a camp counselor in 1991. "We're not reinventing the wheel—we threw out the wheel." The school's founders didn't have any particular allegiance to adventure sports; they just found that climbing, backpacking, and canoeing were a magical fit for these kids, at these ages, when their neurons are exploding in a million directions. "When you're on a rock ledge," Willson says, "there's a sweet spot of arousal and stress that opens you up for adaptive learning. You find new ways of solving problems."

Frances Kuo, the University of Illinois researcher known for her window studies in public housing, has also examined the relationship between ADHD and outdoor activity. Her studies have been small but suggestive. In one experiment, exposure to nature reduced reported symptoms of ADHD in children threefold compared with staying indoors. In another, she had 17 children aged eight to eleven with ADHD walk for 20 minutes with a guide in three different settings: a residential neighborhood, an urban downtown street and a park setting. After the park walk they performed so much better memorizing numbers in backward sequence that the improvement was equal to the difference between having ADHD or not having it, as well as to the difference between not being medicated at all and experiencing the peak effects of common ADHD medication. More recently, a study of 2,000 children in Barcelona found that those who spent more time playing in green spaces were

reported by parents to have somewhat milder symptoms of inatten-
tion and hyperactivity.

In a 2004 paper, Kuo and her colleague, Andrea Faber Taylor, proposed an explanation for how Attention Restoration Theory might
apply. The right prefrontal cortex—the brain's organizing, judging,
task-focusing real estate—is known to be less active in children
with ADHD. If nature allows the right prefrontal cortex to recharge,
it could boost attention in these kids.

ADHD symptoms, it turns out, are somewhat contextual. If
you're the sort of person who thrives on chaos and stimulation like
a lot of extreme athletes, sitting in school all day may well suck out
your soul. But with the rise of industrialism, educators thought all
kids should be in standardized classrooms. "ADHD got its start 150
years ago when compulsory education got started," said Stephen
Hinshaw, a psychologist at the University of California, Berkeley.
"In that sense, you could say it's a social construct."

Not only will exploratory kids feel bored and inadequate in conventional schools, he said, the constrained setting actually makes
their symptoms worse. Maria Montessori went so far as to suggest
in 1920 that middle-schoolers should ditch lecture-based instruc-
tion altogether and head for farm and nature schools where they
can move around and learn by doing. For kids like Zack Smith,
school feels especially stifling and rule-bound; they act up; they
may get moved into an even more restrictive environment, sometimes with chain-link fences, guards, and neurotropic meds that
go beyond ADHD to deal with the ensuing anxiety, depression, and
aggression. Sometimes they end up in trouble, or, as Zack feared
might happen to him, getting "gooned" in the middle of night by
burly strangers who would pack him off to a residential therapeutic
program that may look like Outward Bound in the brochure but end
up feeling like a gulag.

Interestingly, researchers have observed similar patterns in lab rats, who, let's face it, suffer the ultimate cosmic gooning. When Jaak Panksepp, a neuroscientist at Washington State University, restricted the free exploration and play of his young rats, their frontal lobes (which control executive function) failed to grow properly. As adults, they behaved like rat-style sociopaths. "We had the insight that if animals don't play, if there are not sufficient spaces for them to engage, they develop play hunger," said Panksepp. "They have impulse control problems and eventually problems with social interactions."

In contrast, animals given time to play appear to develop deeper and more durable neural hardware. Panksepp's studies show that just thirty-minute play sessions help young rats release brain-growth factors and activate hundreds of genes in the frontal cortex. He points out that while common stimulant medications for ADHD like Ritalin and Adderall may improve attention skills and academic performance in many kids, they do so at the cost of killing the exploration urge, at least temporarily. "We know these are anti-play drugs," he said. "That is clear and unambiguous."

The bigger question is whether the drugs—and all the enforced sedentary behavior—squeeze the adventure impulse out of kids longer-term. Psychologists tend to disagree on this point, but the truth is, no one really knows. It's not a boutique question. Of the 6.4 million diagnosed kids in America, half are taking prescription stimulants, an increase of 28 percent since 2007.

WHEN SOME OF the teens first arrived at SOAR, they were still putting their clothes on backward. They forgot to eat or they couldn't stop. They lashed out in anger and they were easily frustrated. ADHD symptoms appear to express themselves differently in boys and girls. The classic symptoms in boys, which are better understood, are hyperactivity, impulsivity, and distractibility. We all sit somewhere on the continuum of these traits, but people with more severe symptoms appear to have different chemistry in the parts of their brains governing reward, movement, and attention. They may have trouble listening or sitting still, and they get distracted by external stimuli. Easily bored, they tend to be risk-takers, looking for charged activities that help flood their brains with the feel-good neurotransmitters like dopamine, serotonin and norepinephrine, which otherwise get gummed up in the ADHD brain. Kids with ADHD are more likely to suffer head injuries, accidentally ingest poisons, and take street drugs.

Long-standing research suggests that kids like Zack—and indeed, most kids—would be better off in dynamic outdoor learning environments from the very beginning. As Erin Kenny, founder of Cedarsong Nature School on Vashon Island, Washington, has put it, "Children cannot bounce off the walls if we take away the walls."

It's what the man who founded kindergarten had in mind in the first place.

Friedrich Fröbel was born in 1782 near Weimar, in the heart of Germany's ancient forests and lush vales. A student of natural history who came of age under the spell of Romanticism, he was a lover of the French philosopher Jean-Jacques Rousseau. "Everything is good as it comes from the hands of the Author of Nature," Rousseau wrote to Fröbel's delight; "but everything degenerates in the hands of man." In *Émile*, Rousseau made a case for cultivating curiosity and freedom in childhood. This radical notion came to influence every aspect of progressive education. In Fröbel's day, children under the age of seven typically stayed at home or were farmed out to crèches of convenience. Fröbel understood that an education filled with nature and art could instill a lifelong readiness to learn. He believed children would also pick up emotional skills like empathy, as well as a profound sense of the interconnection of all living things.

After working in primary education for a number of years, he started a school for small children in 1837. It was while walking in the woods (walking in the woods!) that he came up with the name: kindergarten. In it, children would absorb the natural world through all their senses. They would grow plants outdoors, exercise, dance, and sing. They would manipulate simple objects like blocks, wooden spheres and colored papers, thus learning, almost despite themselves, the universal laws of geometry, form, physics and design. Fröbel didn't believe in lock-step lesson plans. Children, he said, should be guided largely by their own curiosity and "self-activity." For a while, the idea caught on, but the Prussian government, fearful of instilling free play and, by extension, free atheistic thinking, shut down public kindergartens before Fröbel's death in 1852. Still, his ideas resonated with scores of wealthy, well-connected women who became phenomenally successful international missionaries for the cause. "It was the seed pearl of the modern era, and it was called kindergarten," argues Norman Brosterman in his compelling history, *Inventing Kindergarten*.

Childhood would never quite be the same.

Although, as kindergarten spread to other nations, including America, the concept changed in ways that would have made Fröbel hurl an abacus. He had opposed formal lessons for this age group, and didn't even want alphabetical letters on blocks. But in the late nineteenth century, educators saw the need to prepare children, especially working-class children, for an industrial work life. Kindergarten shifted to more time indoors and the lessons became more programmatic. Despite a brief flirtation with nature schools in the 1960s and 1970s, American kindergarten continued its relentless slouch into sit-down academics.

But Fröbel's naturecentric ideas didn't disappear from Europe. To this day, European kids aren't taught reading and math in earnest until they reach age seven. Germany has more than 1,000 "forest kindergartens" called *Waldkindergärten*, and they are growing in popularity across northern Europe. In these preschools, kids are out in all kinds of weather, playing with natural materials and pretty much having a ball. I'd visited a school called Auchlone in Perth-shire, Scotland, where kids ran happily around climbing trees, play-ing house in twig teepees and hosting a funeral for a dead frog. For snack time, a four-year-old boy helped light a campfire for making popcorn. The school's director, Claire Warden, is a big fan of kids and fire. She's also a proponent of preschoolers handling knives and challenging themselves physically. She'd told me how after a large tree fell over during a storm, the children had spent days sawing and pounding off sharp bits to make it safer to climb upon. This, she explained, launched a typical, nature-based curriculum: the kids improved their manual dexterity, learned about cause and effect, and practiced teamwork.

Warden knows some of these ideas might be shocking to Ameri-can parents and their notions of bubble-wrapped childhoods. "We can't avoid all risk," she'd said. As if on cue, a boy in yellow boots stalked by carrying a junior hacksaw. "Junior hacksaw" would be an oxymoron in America, but here it's another teaching tool. Earlier, I'd seen the same boy with a potato peeler. "What we do is hazard assessment, not risk avoidance," she'd said. "Schools that are bor-ing and not engaging will end up costing parents and taxpayers mil-lions when these children are teens."

Today, a tenth of preschoolers in Scandinavia spend nearly their entire days outside, and another huge percentage spends a signifi-cant portion outdoors. In Finland, outdoor play is integrated into the day throughout primary school to an astonishing degree: it's common for students to be turned out for fifteen minutes out of every hour.

When I was in Finland, I'd asked a sixth-grade teacher named Johanna Peltola why. She was, like many Finns, extremely prag-matic. "When they go outside and get fresh air, they think more clearly," she said. And yet, while American education experts sing the praises of the Finnish school system, celebrating the nation's high spot in global academic standings, they routinely ignore the fresh-air factor. Outdoor play isn't even mentioned in Amanda Ripley's chapter on Finland in *The Smartest Kids in the World*.

Interestingly, Finland reports the same percentage of children diagnosed with ADHD as the United States: about 11 percent, mostly boys. But while most adolescents in the U.S. are taking medication, most in Finland are not.

What Fröbel believed, and the Finns practice, science has affirmed. Nature play enhances at least two activities known to develop children's cognitive and emotional development: exercise and exploratory play. A large meta-analysis of dozens of studies concluded that physical activity in school-age children (4–18) increases performance in a trove of brain matter: perceptual skills, IQ, verbal ability, mathematic ability, academic readiness. The effect was strongest in younger children.

Even more intriguing, researchers at Pennsylvania State University have found that early social skills matter more than academic ones in predicting future success. They followed 750 children for 20 years. The children whose kindergarten teachers rated them as having strong abilities to cooperate, resolve conflicts and listen to others were less likely to later be unemployed, develop substance abuse problems, get arrested, live in public housing, or go on welfare. Germany sponsored an even more ambitious study in the 1970s. There, researchers tracked graduates of 100 kindergartens. Half the programs were play-based (although not necessarily outdoors) and half were academic and instruction-based. The academic students made initial gains; but by grade four they had fallen behind their play-based peers on every scholastic and socioemotional measure used. In a move that would have warmed Fröbel's art stations, Germany reversed its trend toward academic kindergartens.

But, alas, not the United States, where little kids spend more time at their desks than ever. Preschoolers in the United States average just 48 minutes of exercise a day in their schools, even though the recommended level is 2 hours, according to a 2015 paper published in *Pediatrics*. Of that 48 minutes, only 33 minutes is outside. A 2009 study in *Pediatrics* found that 30 percent of third-graders get fewer than 15 minutes of recess a day, and another study found that 39 percent of African-American students had no recess compared to 15 percent of white students.

Parents aren't helping much either. Jane Clark, a University of Maryland professor of kinesiology calls toddlers "container-ized kids" as they spend increasing time in car-seats, high chairs and strollers, and then shift into sedentary media consumption. According to the Outdoor Foundation's research (funded by the U.S. National Park Service and outdoor industry manufacturers), participation in outdoor activities declined among all children, but declined the most—15 percent—among six-to-twelve-year-olds between 2006 and 2014. Those figures include hiking, camping, fishing, cycling, paddling, skateboarding, surfing, wildlife-viewing and other activities, and do not include organized sports.

In 2004, 70 percent of U.S. mothers recalled that they had played freely outside themselves when they were children, yet only 31 percent allowed their children to do the same, despite a drop in crime since then. British children seem equally tethered. Since the 1970s, their children's "radius of activity"—the area around the home where kids are allowed to roam unsupervised—has declined by almost 90 percent, according to a report by the National Trust. While 80 percent of seven- and eight-year-olds walked to school in 1971, by 1990 fewer than 10 percent did so.

In the U.K., two-thirds of schoolchildren do not know acorns come from trees.

———

AT SOAR, MANY students arrive on meds, and many stay on them. At all times, the instructors carry sealed messenger bags full of pharmaceuticals strapped to their torsos like baby marsupials. Though Willson emphasized that SOAR is not a way to get kids off ADHD medication, some do find that they can taper off. Zack's parents told me they were planning to toss his anxiety drugs during his upcoming holiday break, and they expected to lower the dose of his stimulant as well. "The changes in him have been nothing short of miraculous," said his mother, Marlene De Pecol. "Now he's just happy."

If, as the research suggests, outdoor free play is so important to kids' physical and mental health, you might expect to see evidence of illness during this seismic generational shift indoors. And in fact, that's exactly what you see, although it's impossible to draw a direct line to a particular cause. The stats are alarming: Preschoolers are the fastest-growing market for antidepressants in the United States. More than 10,000 American preschoolers are being medicated for ADHD. Teenagers today have five to eight times more clinically significant scores for anxiety and depression compared to young people born in the 1950s. Since 1999, the U.S. suicide rate has increased for nearly all groups, with the steepest rise—200 percent—among girls ten to fourteen years old.

It's well known that childhood obesity rates have tripled and allergy and asthma rates have increased dramatically in the U.S. in the last three decades. According to data from the U.S. Centers for Disease Control and Prevention, nearly one in ten children has a vitamin D deficiency. That's 7.6 million children. And—get this— two-thirds, another 50.8 million, are considered vitamin D "insufficient." We need sunlight for all sorts of bodily processes from regulating our sleep and diurnal rhythms to facilitating proper bone growth to boosting immunity. The problem has gotten so bad that rickets, a disease caused by lack of vitamin D, which had been virtually eradicated, has begun to show up in pockets of the U.K.

and America. The incidence has quadrupled in the two countries' children in the last fifteen years.

When you put little kids in green environments, even if it's just some lawn and shrubbery, they start moving. In schools with conventional urban playgrounds, the boys tend to run around more than the girls. But studies in Sweden show the exercise gap between boys and girls narrows in more naturalistic environments. Nature levels gendered play. The kids in forest kindergartens also tend to get sick less often than their indoor peers, and they host a healthier, more diverse array of microbacteria in their bodies.

Zack Smith is one of the lucky ones. Privileged kids have tons of options, from summer camp to beautifully landscaped schools. But if we really care about children's health, connecting more kids to nature and shaking up early and elementary education, we're going to have to figure it out where most of us actually live and work: in cities, in housing developments and neighborhoods and in public and private schools.

I asked my son, now in seventh grade in D.C., how many minutes of recess he gets per day.

"Recess? We probably haven't had recess in three months."

This was a problem. I called the head of his junior high.

"I know," she said, putting on her appease-the-unhinged-mother voice. "I wish they could go outside more, too, but it's been too muddy, and then the corridors get muddy."

In other words, it was a janitorial problem. In Finland, kids keep their boots by the front door. Maybe schools in the United States don't need more iPads and test prep; maybe they just need more Wellies.

FRANKLY, THERE'S NO time to waste. While active exploration improves learning in both kids and adults, it's adolescents like Zack—whose prefrontal cortex is in the very midst of laying down a lifetime of neurons—who seem to benefit the most. John Green and Meghan Eddy, biobehavioral psychologists at the University of Vermont, exercised some adult and teenaged rats, and then gave them a task to remember how to find food in a maze. The young rats who exercised bested the adults who exercised, doing as well as rats on Ritalin. It seemed the playful, exploratory, and physical adolescent years exist to boost learning in mammals, just as SOAR's Willson intuited. Or, as Green more formally put it, "the adolescent prefrontal cortex is ready to be molded by environmental experience."

So there you have it: the time is now. There's a limited window of opportunity to best launch these kids, and perhaps, in so doing, to safeguard a future of innovative exploration by the very kids who are wired to do it better than anybody else.

The ADHD population is an advance guard. If they can recognize how to better adapt their environments for their brains, there's hope for the rest of us. One thing is clear: human brains seem to grow best when they get some time outside.

After many years languishing in the Formica-filled classrooms of West Hartford, Zack Smith was ready. He and his pals gathered around the fire pit back at camp, bellies full of hamburgers and bread-and-butter pickles. It was very dark out. Tomorrow all fourteen kids would make it the four pitches up Seneca Rocks. A couple of days after that, they'd backpack across the Dolly Sods Wilderness Area, and then they'd visit Stonewall Jackson's grave and read poetry written by the general's sister-in-law. For now, they were tired, if not exactly mellow.

Zack's job for the day was Captain Planet, meaning mighty takerout of trash. Another kid named Max was Scribe. At sixteen, Max was an expeller of colossal farts, and proud of it. "I don't do anything halfway in the outdoors," he said. He had shared with me on the trail that he was also an expert squirrel hunter, climber, and river runner. When he is done with school, he intends to find a job guiding. Beturbaned in a purple bandanna, he opened the group journal and prepared to record notes of the day's events under the narrow beam of a red headlamp.

Zack was lying on his back and looking up at the stars. He was impressed. "We don't have these at home," he said.

PART FIVE

THE CITY IN A GARDEN

12

Nature for the Rest of Us

*If man is not to live by bread alone, what is better worth doing
well than the planting of trees?*

—FREDERICK LAW OLMSTED

In 2008, our species crossed a significant Rubicon of habitat: for the first time, a majority of us lived in cities. We could now be called, as at least one anthropologist has suggested, Metro sapiens. And we're not done. Globally, 2 billion more people will move to cities in the next thirty years. By 2030, there will be 590 million urbanites just in India. China is already half urban; so is Liberia, and the percentage of urbanites in Bangladesh and Kenya quadrupled in recent years.

This momentous urban migration could be a good thing. Cities are often the most creative, wealthiest and most energy-efficient places to live. City dwellers typically experience better sanitation, nutrition, education, gender equality and access to health care, including family planning, than their rural counterparts. The world's growing megacities, though, are not generally the centers of enlightenment that we might hope. In Kinshasa, a city of more than 11 million in the Democratic Republic of the Congo, per capita yearly income is $250. Harvard economist Ed Glaeser has asked how a megacity with such a poor population can "be anything but a hell on earth?" Making cities like Kinshasa livable, he argues, is "the great challenge of our century."

Cities will have to figure out how to cram more people into smaller areas without everyone going literally crazy. Back in 1965, animal behaviorist Paul Leyhausen described what happened to cats in unnaturally crowded environments: they become more aggressive and despotic, turning into a "spiteful mob." In similar conditions, Norway rats forget how to build nests and start eating their own. In confined primates, hormonal systems get goofy and reproduction can plummet. So what about us? Extensive reviews of the medical literature show a 21 percent increase in anxiety disorders, a 39 percent increase in mood disorders and a doubled risk of schizophrenia in city dwellers. Urban living is associated with increased activity in the brain's amygdala—the fear center—and in the perigenual anterior cingulate cortex, a key region for regulating fear and stress.

Meanwhile, a study from Portugal found that people living near industrial "gray space," as opposed to green space, reported "decreased use of coping strategies" and less optimism. That last bit is not trivial; optimism is associated with healthier behaviors, lower triglycerides and mental resilience. We could use some more resilience: globally, depression is responsible for more healthy years lost than any other condition, according to the World Health Organization.

Now that I'd learned about the ways in which being in nature changes our brains for the better, it was time to figure out how to bring the lessons back to where most of us live, in cities. Here are some of the essential take-homes: we all need nearby nature: we benefit cognitively and psychologically from having trees, bodies of water, and green spaces just to look at; we should be smarter about landscaping our schools, hospitals, workplaces and neighborhoods so everyone gains. We need quick incursions to natural areas that engage our senses. Everyone needs access to clean, quiet and safe natural refuges in a city. Short exposures to nature can make us less aggressive, more creative, more civic minded and healthier overall. For warding off depression, let's go with the Finnish recommendation of five hours a month in nature, minimum. But as the poets, neuroscientists and river runners have shown us, we also at times need longer, deeper immersions into wild spaces to recover from severe distress, to imagine our futures and to be our best civilized selves.

Basically, we need hits from a full spectrum of doses of nature. Is it even possible that megaurban habitats could provide them?

To see how an optimist might view our crowded future, I went to a city where the future has arrived: Singapore. It's unusual for being both a city and a country, the only one in the world. More than 5 million Singaporeans, about eight times the population of Washington, D.C., live on an area only four times larger. Singapore is the third-densest country on earth. It is, as planners say, hyperdense. Primatologist Michael Gumert, who teaches at Singapore's Nanyang Technological University, calls the city a human experiment. "It must increase stress in ways we don't fully understand," he told me. "We're undergoing self-domestication," said Gumert. Will Metro sapiens evolve fast enough to adapt?

When I pictured Singapore, I thought of the ban on chewing gum and public spitting, enforced by arcane punishment like caning. Ridiculed globally for these policies, the city-state brings to mind Nanny McPhee-meets-the-death penalty. But then I heard about Singapore's green walls, its lavish parks and vertical farming, how it is sometimes considered the top "biophilic city" in the world. Flying in, it's immediately obvious that this is a verdant megalopolis, with huge housing blocks interspersed with lush greenery. The roadway from the airport is bordered by palm trees, flowering shrubs and a spreading green canopy. This isn't surprising in a tropical island, but then I learned that this part of the city rests on reclaimed damaged land. Massive deforestation had left the place barren of nutrients. Every one of those trees and shrubs was planted, on imported soil. Like an insecure diva, the city wants you to notice. My hotel and many other buildings downtown looked like chia plants, every few stories and sometimes entire walls sprouting cascading layers of plants. "You can wake up and start grazing!" joked my cabdriver as he dropped me off.

I thought a good place to start diving into the country's nature ethic might be the world-class, 155-year-old Singapore Botanic Garden, which is large, open nineteen hours a day, and free. A new UNESCO World Heritage Site, it's also the headquarters for the country's powerful national parks agency. I ducked out of a downpour and into the administrative building, where I was met by bespectacled Yeo Meng Tong, the affable director of parks development. In most nations, the parks departments are small, underfunded and scrappy. But this country spends 200 million Singapore dollars per year "to develop scenery," as Yeo put it. That equals .6 percent of the national budget, five times the share the National Park Service gets from the U.S. federal budget. No wonder he was smiling.

Yeo told me he was born in 1963, two years before the former British colony cleaved from Malaysia. Under the fifty-year leadership of one ruling party—and mostly one man, the late Prime Minister Lee Kuan Yew—Singapore grew into the third-most-successful economy in the world, ranked higher than the United States on GDP per capita, educational attainment, standard of living and life expectancy. Its accomplishments are all the more impressive given that the place had virtually no exportable natural resources, little room to expand, and a surging population made up of a potentially volatile mix of ethnicities.

Lee Kuan Yew—or LKY, as he's fondly known—planted a public

tree in a traffic circle soon after he took office, setting off what would become a personal obsession. Singapore was soon importing thousands of trees and hiring small armies of arborists and horticulturalists. He launched a "garden city" plan that later morphed into a more ambitious "city in a garden" vision. In his memoir, he writes: "After independence, I searched for some dramatic way to distinguish ourselves from other Third World countries. I settled for a clean and green Singapore. One arm of my strategy was to make Singapore into an oasis in Southeast Asia. . . ."

As Yeo proudly told me, if you add up the forest preserves, the pocket parks, undeveloped land and the manicured street trees, half of Singapore's 276 square miles is under some sort of green cover. "We try to create more green in every inch of space we can find," he said. The city day-lighted and landscaped its once-utilitarian canals, adding paths, so it now offers 300 kilometers of green corridors that connect the many parks. When a new development goes in, the builders must figure out how to more than replace the nature it displaced, by making green roofs, integrated gardens, parks over parking lots, and so on. The government will help fund the extra costs. I visited several mesmerizing structures, including the "world's largest vertical garden," a twenty-four-story condo tower whose entire west face was covered by 23,000 *Thunbergia grandiflora* vines. The effect was a little bit Body Snatchers: the wall was alive! The builders calculate a 15 to 30 percent savings in energy use from better insulation and reduced air-conditioning, a big deal on a tropical island on a warming planet.

Because of these policies, the country's percentage of green space is actually increasing. Even while the population grew by some 2 million between 1986 and 2007, the percentage of green space expanded from 36 to 47 percent. By contrast, my city, Washington, D.C., has experienced the opposite, along with most places on the planet: only 36 percent of the overall tree canopy remains, a decrease from 50 percent in 1950. Singapore is a remarkable model of what's possible when green gets coded into a city's DNA. Furthermore, "we try to achieve a goal that 80 percent of people live within 400 meters of green space," Yeo said. "We're pretty close. Now we're at 70 percent."

Yeo bounded outside, where the rain had ceased, to show me the garden's heritage trees. One, a sprawling, 150-year-old native Tembusu tree, is so beloved that it graces the five-dollar bill. A long, horizontal branch as thick as a barrel thrusts out from the trunk not far above the ground. "This is a sentimental tree for many Singaporeans because children grow up climbing it on the family outing," he said. "And then they hang out there with their friends, and it becomes a dating tree, then a proposal tree, and then people take their wedding pictures here!"

"Was your wedding picture here?" I asked.

"Yes!"

IT ALL SOUNDED good, but like much in Singapore, the nature love was well packaged, ready-made for brochures and airport posters. Were all the nice parks and green-carpeted buildings the ones the tourists and investors see? Was this a Potemkin paradise? To examine the reach of nature into the lives of real people, I visited a community hospital, Khoo Teck Puat. It's not close to the center of the city, and it's not used much clinically by foreigners or expats. But it's known as a new and successful example of simple biophilic design. I have to say, it was gobsmackingly nice, especially for a hospital. Many rooms faced the inner, luxuriant garden courtyard, dense with trees and shrubs specifically selected to attract birds and butterflies. Outside sat a sizable pond, a medicinal herb garden and a walking path. Artificial mini islands floated in the pond to attract egrets. The overall site employed a conscious design for biodiversity: endangered fish swam in a little watercourse that wove through the garden. Sadly, this is about the only habitat they have left.

Plants draped over balconies on each floor, giving the impression the building had just risen from the jungle floor, adding to the Shangri-La effect. "We call it the hospital in a garden," said chief gardener Rosalind Tan, who is sometimes called Madame Butterfly, as we walked by a blooming hibiscus, popular with the tiny golden sunbird. "We know from practical experience that people enjoy greenery and we try to create a healing environment for patients so they can have lower blood pressure and be in a better condition to see a doctor."

We walked through the spotless ICU, where every patient has a view of trees out six-foot windows. At many points, corridors and landings open up to the outdoors. I noticed none of the usual antiseptic hospital smell, despite the place having one of the lowest hospital-acquired infection rates in the country, according to Tan. I was reminded of a 2012 study from a Portland, Oregon, hospital showing that rooms with better ventilation from outside garnered more diverse bacterial profiles and fewer "bad" bacteria. Tan next showed me the organic vegetable garden on the roof, which is mostly tended by locals who enjoy gardening. Patients eat some of the produce, and some is sold in a farmers' market. She plucked a few long purple and green leaves off a rhoeo oyster plant and gave them to me to make a tea. "Our signature drink, full of antioxidents," she said. "Good for cooling."

I went back to my chia-plant hotel and brewed some. Then, newly cooled, I headed out again. Everyone told me that before I left Singapore, I had to see the Gardens by the Bay. This is a huge, showy billion-dollar attraction on the newly reclaimed waterfront land. A "premier urban recreation space," it consists of numerous outdoor gardens and two ginormous horticultural greenhouses. Typically, such conservatories have to be heated; here, they have to be cooled. They showcase biozones from temperate climates, including cloud forests, Mediterranean olive groves and the California chaparral.

But the park's piece de resistance is a grove of eighteen Supertrees that are entirely fake. Better than the real thing, they soar between 80 and 160 feet into the sky like giant skeletal golf tees. A narrow walkway snakes through the canopy of a few of them so that you can view the city skyline unencumbered and then eat high-end egg rolls on cowhide cushions at the penthouse restaurant. The structures collect and sprinkle rainwater on the (real, but planted) vines and bromeliads growing on them. They collect solar power in panels, and, best yet, they convert that electricity into an evening light extravaganza.

Recovering from the egg rolls, I settled onto the finely clipped lawn below, surrounded by couples and small children running around on the family outing. The sky grew dark, and the first notes of an electronic symphony began. Suddenly, the trees erupted in colorful neon bursts that kept perfect time with the symphony. The Led Zeppelin stoner laser show has nothing on this. I felt an emotion not dissimilar to what I experienced in the canyons of Bluff, Utah. I felt the stirrings of awe.

This was nature in the Future City, a mix of metaphor, technology and evolutionary impulse. It embodies what the writer and digital pioneer Sue Thomas calls "technobiophilia." Who's to say what real nature is anymore anyway? The human hand underlies all of the world's ecosystems now. Singapore just represents the extreme end of constructed nature. It still pushes many of our neurological buttons for grass, green, blue, safety, beauty, play, visual interest, wonder. Could I find it truly satisfying? Could any of us who have spent time in wilderness? In a word, no. It wasn't unpredictable and therefore couldn't be interesting for long; it didn't stay novel or fulfill the Kaplans' quotient of being mysterious or escapist enough. But I looked at these children, and their young parents, and I realized that most of them had probably never seen a much wilder nature, and they didn't miss what they didn't know. If this isn't an argument for conserving wilderness and making sure people experience it, I don't know what is.

Heading out of the park, a fragile sliver of hazy moon hung in the southern sky.

I hadn't noticed it at all.

I TOOK AWAY two big lessons from Singapore. For greenery to truly seep into all neighborhoods, there needs to be a strong governing vision. Second, urban nature will serve us best when it's allowed to be a little bit wild, at least in spots. I couldn't help but wonder if cities had something better to offer in the awe department. Real nature, the kind we evolved in, incorporates entropy, blood, high winds, a beating, pulsing geophony. In Singapore, nature more or less looked like nature, but it didn't sound like nature. It didn't act like nature. Where was the possibility of all that Darwinian tooth and claw?

Celebrating living trees instead of fake trees seemed like a logical first step. In fact, trees might be our single best tool for urban salvation. City dwellers get most excited about two natural features: water and trees. Now fans can even write emails to trees in Melbourne ("As I was leaving St. Mary's College today I was struck, not by a branch, but by your radiant beauty. You must get these messages all the time. You're such an attractive tree." The trees, which are tagged with individual identification numbers in St. Mary's Park, sometimes write back via the park crew).

My man Olmsted understood this devotion. In his principles for park design, he thought no features should stand out as too distracting or spectacular. There should be no flamboyant flower beds and only a minimal amount of overt architecture. The magic formula: generous meadows loosely defined by trees. Winding pathways leading to mystery, flirtatiously half concealed by trees. Trees, trees, trees. They were so important to the Olmsted schema that he ordered no fewer than 300,000 of them for Central Park's 800 acres, effectively freaking out his budgetary overlords. There were so many trees and shrubs that Calvert Vaux had to recruit a small team of family and friends to fill in the master drawing with tiny green spots. This was pixelation, circa 1858.

Urban trees provide not just aesthetic pleasure but concrete health benefits. Although certain species of trees can worsen asthma through pollen and other compounds, taken as a whole they generally improve people's physiology in several important ways. Public officials perhaps didn't fully appreciate this until a rather astounding study was published in 2013. Geoffrey Donovan, an urban forester with the U.S. Forest Service, spotted an intriguing natural experiment: a pesky scourge called the emerald ash borer, a "phloem feeder," landed on our shores in about 2002, whereupon it decimated 100 million ash trees throughout the Midwest and Northeast. Gone, poof. Donovan decided to see if there was any relationship between the treepocalypse and the incidence of cardiovascular disease in humans.

Donovan was already aware of some seminal European studies looking at human stress, illnesses and loosely defined "green space" in cities. And there were other studies, including Richard Mitchell's work in Scotland, showing lower mortality rates near urban parks. While Mitchell's research revealed a big health boost to poor people, Donovan's work showed the sudden tree blight had a bigger impact on wealthier neighborhoods, probably because those had the most trees to lose. Overall, the counties that were hit by the borer suffered 15,000 additional deaths from cardiovascular disease and 6,000 more from lower respiratory disease. Those figures represent a sizable 10 percent increase in expected mortality. It's hard to say whether the deaths were caused by worsened air quality or changes in stress brought on by not having the tall, green, comforting trees to look at, or both. If trees can move us so powerfully in their meta-

phoric reach, as the veterans on the Salmon felt, then perhaps looking at sick or dead trees is in itself stressful.

Toronto takes its 10 million trees very seriously, valuing its urban
forest at $7 billion. A recent study there showed the higher a neigh-
borhood's tree density, the lower the incidence of heart and metabolic disease. Putting it into raw economic perspective, the health
boost in those living on blocks with about 11 more trees than average was equivalent to a $20,000 gain in median income. Lucky resi-
dents were rich in trees.

Every tree helps. As the founding nature/brain researcher
Rachel Kaplan told me, "nature doesn't have to be pervasive. One
tree is an awful lot better than no tree." But more trees are best.
The city of Washington, D.C., and partner nonprofits have been
trying to plant at least 8,600 trees a year in an effort to increase the
street canopy to 40 percent in the next two decades. New York City
recently completed a wildly ambitious campaign to plant a million
trees, and Los Angeles, Shanghai, Denver and Dubai are in the
middle of similar ones.

Trees are considered a critical part of the global carbon storage
solution, the heat-island solution and the urban air-quality solution.

It's a tall order, but they stand at the ready.

Epilogue

But are not exercise and the open air within the reach of us all?

—WALT WHITMAN

f there's one major theme of this book, it's that the benefits of nature work along a dose curve. Tim Beatley, who runs the Biophilic Cities Project at the University of Virginia, promotes a concept called the nature pyramid. It's a recommended menu for getting the nature humans need, and I think it's a genius idea. It also happens to mirror the structure of this book, from quick doses of nearby nature to longer spells in wild places. Inspired by the ubiquitous food pyramid, Beatley places at the base the daily interactions with nearby nature that help us destress, find focus and lighten our mental fatigue. These are the birds and trees and fountains in our neighborhoods, our pets and our house plants, public and private architecture that allow for daylight, fresh air and patches of blue sky and naturalistic landscaping. These are our daily vegetables, and Singapore, laser lights and all, has it nailed. We should all be so lucky.

Moving up the pyramid are weekly outings to parks and waterways, places where the sounds and hassles of the city recede, places that we should aim to imbibe at least an hour or so a week in the Finnish fashion. These might include wilder, bigger city parks if we're lucky, or regional parks that we can travel to fairly easily.

Moving up higher still are the places that take more effort to get to: the monthly excursions to forests or other restful, escapist natural areas along the lines of what Japan's Qing Li recommends—a weekend per month—for our immune systems.

At the very pinnacle are the rare but essential doses of wilderness, which Beatley and scientists like Utah's David Strayer think we need yearly or biyearly, in intense multiday bursts. As we've seen, these trips can rearrange our very core, catalyzing our hopes and dreams, filling us with awe and human connection and offering a reassurance of our place in the universe. There may be particular times when wilderness experience can be most helpful to us, such as during the identity-forming roller coaster of adolescence or following grief or trauma.

The more we recognize these innate human needs, the more we stand to gain. I'd love to see more wilderness therapy, more kids in summer camp and on nature field trips and on scouting expeditions and on quests of one kind or another, and more opportunities for city populations in general to touch the wild. We all need a regular check-in for personal introspection, goal-setting and spiritual reflection. Best to turn the phone off.

Distilling what I learned, I came up with a kind of ultrasimple coda: Go outside, often, sometimes in wild places. Bring friends or not. Breathe.

According to Beatley, there's cause for hope. Cities around the world are undertaking projects large and small to integrate a range of natural elements into everyday life, and they're seeing huge payback, from New York's High Line to the opening up that we saw of South Korea's Cheonggyecheon River. When cities become greener, it makes not only people more resilient but the cities themselves.

They can better handle extremes of moisture and temperature; they rebound more quickly from natural disasters and they provide refugia for disappearing species from bees to butterflies to birds and fish.

Since our brains especially love water, it makes sense to put it at the heart of these projects. Thirty-two miles of the Los Angeles River are being transformed from a concrete-lined eyesore into a biological and recreational corridor. Copenhagen now has several safe swimming areas in the harbor. People swim in organized events from San Francisco's Baker Beach to Alcatraz. Washington, D.C.'s Anacostia River, once a forgotten, crime-ridden excuse for sewage, now hosts Friday Night Fishing for families and canoe trips for schoolchildren. But try topping this: Wellington, New Zealand, offers a public snorkel trail. Such places exemplify, said Beatley, "cities of awe." But the challenge remains to make "blue space," whether awesome or merely restorative, accessible to everyone.

We still have a long way to go. You can see poverty from space. My own city, D.C., has a clear "tree line" that can be seen in satellite photos analyzed by the *Washington Post*. To the west of that line, in the affluent Northwest quadrant, the streets glow green from above. To the east, where 40 percent of residents live in low-income neighborhoods, the area looks flat and gray. The picture is hardly unique, and this inequality is our essential conundrum as we move toward increasingly urban habitats.

Olmsted understood that throughout history—from the ancient Persians to the English gentry, whose manicured hunting grounds first inspired city parks—the rich always got to enjoy restful glades and pastures. Olmsted wanted to break that pattern fundamentally. Not only did he want people to heal in parks; he wanted all people to have the chance. In the 1870s, he actually posted notices in tenements and sent circulars to all the doctors in New York City with directions to Central Park and Prospect Park; the posters included a description of natural destinations to aid convalescents.

Why shouldn't doctors prescribe time outside to their patients?

It's taken nearly 150 years for Olmsted's idea to gain some traction. There aren't many doctors sending their urban patients to the park, but there are a few. Nooshin Razani, a pediatrician at Children's Hospital in Oakland, California, has forged a partnership with local parks so inner-city kids can get to them more easily and more often. Like Razani, Robert Zarr, a pediatrician at Unity Healthcare in Washington, D.C., saw that conventional approaches weren't serving his underprivileged patients. Many were suffering from obesity, diabetes, depression, anxiety and asthma.

"This is a no-brainer," he said. "Parks are free. They are an incredible resource not being used. We just need to connect people to them."

Health care is only a piece of the solution. The access-to-nature movement also ideally needs to grow out of schools, churches, workspaces, neighborhood associations and cities as a whole. And it won't happen unless we acknowledge more consciously our need for nature. As I've learned through the course of reporting this book, we profoundly undervalue that need. You can see it when we cut recess and outdoor play for kids, when we design buildings and neighborhoods that cut off light, space and fresh air, when we stay inside instead of making the effort to get out. The wealthier you are, the more likely you are to satisfy your nature neurons, but it's often a subconscious fulfillment met by exclusive neighborhoods and restorative vacations. Until we all fully acknowledge the need for nature that's driving some of our behavior, we won't work to make it available for everyone.

I'm heartened by the small bursts of activism taking place in communities throughout the country, whether through fun and innovative groups like Outdoor Afro, GirlTrek, CityKids, Nature Bridge, the Children & Nature Network and dozens of others. Adventure playgrounds—complete with mud puddles and you-build-it twiggy forts are springing up in places like Houston, Texas, and Governors

Island, New York. So-called "tactical urbanists" are installing pop-up parks and guerilla gardens on city streets. Increasingly, organizations, public agencies and institutions are working hard to get people, including me, into the thin ribbons of blue-green that still weave through our urban habitats. It's no longer enough to save wild places from people—now groups are saving them *for* people. The Nature Conservancy, known for preserving important ecosystems and habitats, created a new Human Dimensions Program (HDP), an initiative to bring human well-being considerations into conservation practice. The U.S. National Park Service introduced a major Healthy Parks, Healthy People initiative, specifically geared toward making parks more attractive to diverse populations for both the health of the parks (so they'll be used) and the health of people. "In the past we tended to encourage visitors to come to the parks and have fun and learn something and be safe," Diana Allen, chief of the service's Office of Public Health, told me. "Now we say come have fun and be healthy. That's huge."

If we value how important access to parks is for neighborhood well-being, then we need to measure it. The nonprofit Trust for Public Land recently compiled a helpful "ParkScore" index, ranking every major U.S. city by the proportion of residents living within a 10-minute walk of a park. Minneapolis ranked first (no wonder they're so happy there!), with 86.5 percent success. I was surprised to see Washington, D.C., ranked third, at 80 percent, if you include public lawns like the National Mall.

I'll admit, I'm still struggling to make peace with my own migration to the city, but my mood, along with my habits, are getting better. Since starting this book, I've changed the way I walk around, seeking out the routes with more trees. I go to parks a lot, and I walk in them often. I make my kids come with me. We make an effort to listen to the birds, to look at the fractal patterns in nature, to watch the creeks flowing. I still shake my fists at the planes, but I also enjoy getting on them to go somewhere more wild.

This winter, we had a blizzard big enough that it stopped virtually all mechanized air and street traffic for a couple of days. The
deer took back the streets, bounding through the city in the snow.
People frolicked in the streets too, sledding down boulevards, doing
handstands, stomping around between shoveling sessions. When
the sun came out, my husband and I laced on some old ski boots
and schussed down to the canal path. We were about the only people
down there.

"It's so quiet!" I said.

"We could be in Yellowstone!" he said.

We heard a few titmouses and cedar waxwings.

On our way back home, we passed an old Italian woman surveying the shoveling work of some teenagers. She said, "So pretty out!"
I said, "No planes!" and her expression took on a revelatory look and
she laughed and said, "Brava! No planes!"

Then we skied back toward the house and I cheered on a man
who was almost done shoveling his epically buried car. We ran into
some neighbors we hadn't seen in two years and found out one had
been undergoing cancer treatment. We talked for half an hour. We
came upon a pack of enterprising boys and hired them to shovel our
driveway. When they finished, they came in to watch the last plays
of a Broncos game along with our next-door neighbor, who brought
snacks. "It's like a neighborhood again," he said.

It was still the city, but it had been, if not taken over by natural
forces, at least temporarily matched by them. Nature asserted itself
and the city watched, and played.

ACKNOWLEDGMENTS

I owe huge thanks to my editors at *Outside* magazine—Elizabeth Hightower, Michael Roberts and Chris Keyes—who sent me down this leafy errand in the first place, and to my editor Rob Kunzig at *National Geographic* for helping me complete the journey. The photographs of Lucas Foglia inspired me and I'm grateful to include them in these pages.

Proving that time looking at and thinking about nature makes people generous, numerous researchers, too many to list, opened up to me their labs, minds, and field sites and withstood my endless questions, sometimes in a language not their first. But I need to call out a few for uncommon offerings of time and expertise: Juyoung Lee, David Strayer, Adam Gazzaley, Art Kramer, Liisa Tyrvainen, Kalevi Korpela, Deltcho Valtchanov, Jenny Roe, George Mitchell, Ulrika Stigsdotter, Patrik Grahn, Matilda Van Den Bosch, Greg Bratman, Marc Berman, Derrick Taff and his team, and Tan Le. Special thanks to my Korean translator, Sepial Shim. I look forward to following their work.

Private and institutional support were critical to this project. I'm grateful to Virginia Jordan and Bill and Elaine French for their support. Thanks to Brooke Hecht, Curt Meine and Gavin Van Horn at the Center for Humans and Nature, my fiscal sponsor and a great think tank. Thanks as well to Melissa Perry and the Department of Environmental and Occupational Health at George Washington University for sponsoring my professorial lectureship, which provided free and extensive library access.

Writing a book didn't just happen in neat nine-to-five increments in my office. I often got away for intense bursts, and many people helped me out and sometimes kept me company. I was fortunate to receive a two-week residency at Mesa Refuge in Point Reyes with Sarah Chang and Zahir Janmohammed, who kept me in good supply of Korean BBQ and chai. Thanks for that magical place to Peter Barnes, Susan Page Tillett and Patricia Duncan. Also thanks to the good people of the Virginia Center for the Creative Arts. My brother and sister-in-law Jamie and Wendy Friar also let me hole up in their basement for a few days to write. Thanks to Michelle Nijhuis for cavorting with me between tacos and writing spurts in the Sonoran Desert and to Margaret Nomentana for supplying a lake in Maine, a couple of lobsters, and some loving childcare. I couldn't have escaped my parenting duties without other helpful assists from Penny Williams, world's greatest mother-in-law, Rachel Baranowski and Allison Frisch. Kate Sheridan and Danielle Roth ably helped with some research and fact-checking.

I'm very grateful to my D.C. writing pals and peers, Josh Horwitz, Juliet Eilperin, David Grinspoon, Eric Weiner, Tim Zimmermann, Jacki Lyden, Maarten Troost, Margaret Talbot, Alex Zapruder and Hanna Rosin. You all made my transition to D.C. infinitely better, and you inspire me with your smarts and talents. My old writing pals in Boulder continued to offer support, read large portions of the manuscript and entertained me with funny tales of the boho town I left. Thank you Hannah Nordhaus, Hillary Rosner, Melanie Warner and, especially, the talented Lisa Jones, who in addition to being my partner in many adventures is also my dear sister-in-law. Also thanks to D.C. buddies Eliza McGraw, Kim Larson, Donna Oetzel, Margaret Reitano, Melissa Boasberg, Will and Erica Shafroth, Kirk Johnson and Chase DeForest, and to farther- flung friends Julie Frieder and Ann Vileisis, who all offered help as sounding boards. Sometimes we even got out in nature.

Flora Lichtman provided graphics inspiration. I may not be a fan of all things virtual, but I'm fortunate to have a cyber coterie of stellar science writers who offer commiseration, blow me away with their talents and kick my ass when needed: Christie Aschwanden, Bruce Barcott, Maryn McKenna, Seth Mnookin, David Dobbs, Deborah Blum, Elizabeth Royte and Karen Coates. Our coffee mug says WTMFB. I'm fortunate to have all of you in my life.

Several other people read all or parts of the manuscript and offered critical advice. Big gratitude for the helpful insights and occasional fist-pumping of Amanda Little and Jay Heinrichs. Both of them are prodigiously talented. Thanks to my nature-loving and earth-moving agent Molly Friedrich and to the amazing team at W. W. Norton—especially Jill Bialosky, Maria Rogers, Erin Sinesky Lovett, and Steve Colca—and, for epic copyediting, Fred Wiemer.

There would be no fun in the natural world (or anywhere else) without my game and loving family, both extended, stepped, halved, in-lawed and nuclear. This book is really for and about you, John Williams, Jamie Williams, Ben Williams and Annabel Williams. There's no nature like shared nature.

NOTES

INTRODUCTION: THE CORDIAL AIR

1   Title, "The Cordial Air," from Ralph Waldo Emerson's essay, *Nature*, first published in 1836. "In good health, the air is a cordial of incredible virtue."

1   "May your trails": From Edward Abbey, *Desert Solitaire: A Season in the Wilderness* (Tucson: University of Arizona Press 1988), preface.

2   About the MacKerron study quoted: It's worth pointing out that MacKerron controlled for lots of variables, such as weather, companionship, etc., and he also was able to factor in the vacation effect by looking only at responses given during weekends and national holidays, when presumably most people were not working. In other words, people weren't just reporting feeling happier because they were off work whenever they were in nature. Everyone was off work, so the playing field was more level. From George Mackerron and Susana Mourato, "Happiness Is Greater in Natural Environments," *Global Environmental Change*, vol. 23, no. 5 (Oct. 2013): p. 992.

3   As Nisbet rather dejectedly concluded: Elizabeth K. Nisbet and John M. Zelenski, "Underestimating Nearby Nature Affective Forecasting Errors Obscure the Happy Path to Sustainability," *Psychological Science*, vol. 22, no. 9 (2011): pp. 1101–6.

4   We check our phones 1,500 times a week: Based on a survey in the U.K. by a marketing agency, Tecmark.http://www.dailymail.co.uk/sciencetech/article-2783677/How-YOU-look-phone-The-average-user-picks-device-1-500-times-day.html, accessed May 26, 2015.

4   iPhone users vs. Android users: From an Experian marketing survey, written about here http://www.experian.com/blogs/marketing-forward/2013/05/28/americans-spend-58-minutes-a-day-on-their-smartphones/, accessed May 27, 2015.

4   Regarding children spending little time outside: Only about 10 percent say they

are spending time outdoors every day, according to a Nature Conservancy poll, http://www.nature.org/newsfeatures/kids-in-nature/kids-in-nature-poll.xml.

4   "Tired, nerve shaken, over-civilized people": John Muir, *Our National Parks* (New York: Houghton, Mifflin, 1901), p. 1.

4   "pestiferous little gratifications": From Mose Velsor (Walt Whitman), "Manly Health and Training, with Off-Hand Hints Toward Their Conditions," ed. Zachary Turpin, *Walt Whitman Quarterly Review* 33 (2016): p. 289.

4   Wordsworth lines: from *The Prelude*, 1805.

5   Beethoven's tree: Cited in Eric Wiener, *The Geography of Genius* (New York: Simon & Schuster, 2016), p. 235. The Beethoven quote is from his letter to Therese Malfatti in 1808.

5   For more on prospect and refuge theories of human habitat preference, see Jay Appleton, *The Experience of Landscape* (London: John Wiley, 1975) and Gordon Orians, *Snakes, Sunrises and Shakespeare* (Chicago: University of Chicago Press, 2014).

6   We've become arguably more irritable, less sociable, more narcissistic: see studies by Clifford Nass, including Roy Pea et al., "Media Use, Face-to-face Communication, Media Multitasking, and Social Well-Being Among 8-to-12-Year-Old Girls," *Developmental Psychology*, vol. 48, no. 2 (2012): p. 327 ff. On nature deficit disorder, see Richard Louv, *Last Child in the Woods* (New York: Workman Publishing, 2005).

11   On Taksim Gezi Park, see Sebnem Arsu and Ceylan Yeginsu, "Turkish Leader Offers Referendum on Park at Center of Protests," *New York Times*, June 13, 2013. http://www.nytimes.com/2013/06/13/world/europe/taksim-square-protests-istanbul-turkey.html?_r=0, accessed July 2, 2015.

12   Olmsted quote: can be found in Witold Rybyznski, *A Clearing in the Distance: Frederick Law Olmsted and the Nineteenth Century*, Kindle location 4406.

CHAPTER 1: THE BIOPHILIA EFFECT

Portions of this chapter originally appeared in Florence Williams, "Take Two Hours of Pine Forest and Call Me in the Morning," *Outside*, Nov. 2012, published online Nov. 28, 2012.

17   "In short, the brain evolved in a biocentric world": Edward O. Wilson, *The Biophilia Hypothesis* (Washington, DC: Island Press, 1993), p. 32.

17   "There is nothing you can see that is not a flower": Matsuo Basho quoted in Margaret D. McGee, *Haiku—The Sacred Art: A Spiritual Practice in Three Lines* (Woodstock, VT: Sky Paths Publishing, 2009), p. 32.

19   With the largest concentration of giant trees: Miyazaki from the book *Designing Our Future: Local Perspectives on Bioproduction, Ecosystems and Humanity*, ed. Mitsuru Osaki: Okutama Town designated in 2008, pp. 409–10.

19   68 percent of the country's land mass: Qing Li. "Effect of Forest Bathing Trips on Human Immune Function," *Environmental Health and Preventive Medicine*, vol. 15, no. 1 (2010): pp. 9–17.

19   one hundred Forestry Therapy sites within ten years: Yoshifumi Miyazaki, "Science of Nature Therapy," p. 8, http://www.fc.chiba-u.jp/research/miyazaki/assets/images/natural%20therapy(07.06)_e.pdf, accessed June 2015.

20   In addition to those: "Suicide in Japan," *Japan Today*, Jan. 18, 2011.

20   commuting hell: Eric Goldschein, "Take a Look at Why the Tokyo Metro Is Known as 'Commuter Hell,'" *Business Insider*, Jan. 11, 2012; and Ronald E. Yates, "Tokyoites Rush to 'Commuting Hell'" *Chicago Tribune*, Oct. 28, 1990.

21   Erich Fromm, who described it in 1973: Fromm quote from *The Anatomy of Human Destructiveness* (New York: Holt, Rinehart & Winston, 1973), p. 366. Cited in Stephen R. Kellert, *Kinship to Mastery: Biophilia in Human Evolution and Development* (Washington, D.C.: Island Press, 1997).

21   Wilson distills the idea more precisely: Stephen R. Kellert and Edward O. Wilson. *The Biophilia Hypothesis* (Washington, D.C.: Island Press, 1995), p. 416.

23   As Miyazaki explained it: See Yoshifumi Miyazaki, "Science of Nature Therapy" (above) and Juyoung Lee et al., "Nature Therapy and Preventive Medicine," in *Public Health—Social and Behavioral Health*, ed. Jay Maddock (Rijeka, Croatia: InTech, 2012); and Miyazaki et al. "Preventive Medical Effects of Nature Therapy," *Nihon eiseigaku zasshi/Japanese Journal of Hygiene*, vol. 66, no. 4 (2011): pp. 651–56.

25   We suffer the consequences: Sandor Szabo, Yvette Tache, and Arpad Somogyi, "The Legacy of Hans Selye and the Origins of Stress Research: A Retrospective 75 Years After His Landmark Brief 'Letter' to the Editor of Nature," *Stress*, vol. 15, no. 5 (2012): pp. 472–78.

25   heart disease, metabolic disease, dementia and depression: Esther M. Friedman et al., "Social Strain and Cortisol Regulation in Midlife in the US," *Social Science & Medicine*, vol. 74, no. 4 (2012): pp. 607–15.

27   The brains-on-built-environment: Roger S. Ulrich et al., "Stress Recovery During Exposure to Natural and Urban Environments," *Journal of Environmental Psychology*, vol. 11: 201–30.

28   But Li found similar results with NK cells: Qing Li et al., "Effect of Phytoncide from Trees on Human Natural Killer Cell Function." *International Journal of Immunopathology and Pharmacology*, vol. 22, no. 4 (2009): pp. 951–59.

CHAPTER 2: HOW MANY NEUROSCIENTISTS DOES IT TAKE
TO FIND A STINKING MILKVETCH?

33   "We used to wait": Arcade Fire, "We Used to Wait," from *The Suburbs*, 2010.

37   a 50 percent improvement in creativity: The four-day wilderness pilot study

is R.A. Atchley et al., "Creativity in the Wild: Improving Creative Reasoning Through Immersion in Natural Settings," *PLoS ONE*, vol. 7, no. 12 (2012), published online, e51474.

42   "Every one knows": William James, *The Principles of Psychology* (Chicago: Henry Holt/ Encyclopedia Britannica, 1991), p. 261.

42   "My experience is what I": James, p. 260.

42   "spiritual alertness of the most vital description": William James quote from the biographical note in James, p. vi.

43   "I am away from the office": From the Twitter feed of Shit Academics Say, May 13, 2015, 9:41 P.M., https://twitter.com/AcademicsSay.

43   For perspective, it takes: The brain's processing speed is about 120 bits per second, from Daniel Levitin, *The Organized Mind: Thinking Straight in the Age of Information Overload* (New York: Dutton, 2014), p. 7.

43   Moreover, task-switching: Task-switching burns up oxygenated glucose . . . Levitin, p. 98.

46   "The average American": Levitin, p. 12.

48   "employs the mind": Olmsted's 1865 Report to the Congress of the State of California as quoted in Roger S. Ulrich et al., "Stress Recovery During Exposure to Natural and Urban Environments," *Journal of Environmental Psychology*, vol. 11, no. 3 (1991): p. 206.

49   partly "recovered": The Kaplan/Berman cognitive study: Berman et al., "The Cognitive Benefits of Interacting with Nature," *Psychological Science,* vol. 19, no. 12 (2008): pp. 1207–12.

53   At least one MRI study: The MRI study showing increased activation in the insula and anterior cingulate is Tae-Hoon Kim et al., "Human Brain Activation in Response to Visual Stimulation with Rural and Urban Scenery Pictures: A Functional Magnetic Resonance Imaging Study," *Science of the Total Environment,* vol. 408, no. 12 (2010): pp. 2600–2607.

CHAPTER 3: THE SMELL OF SURVIVAL

Some of the material in this chapter appeared in different form in Florence Williams, "This is Your Brain on Nature," *National Geographic,* January 2016.

59   "I can't begin to count": Euny Hong, *The Birth of Korean Cool: How One Nation Is Conquering the World Through Pop Culture* (New York: Picador, 2014): p. 61.

62   South Korea then had a lower GDP: Hong, p. 2.

62   One-third of Koreans were homeless: Daniel Tudor, *Korea: The Impossible Country* (North Clarendon, VT: Tuttle Publishing, 2013), Kindle location 171.

62   "that quality of air": From *Essays of Travel* (London: Chatto & Windus, 1905), p. 170, http://www.archive.org/stream/e00ssaysoftravelstevrich#page/n7/mode /2up, accessed 6/17/15.

62   "The piny sweetness": From "Pan in America" and cited in Tianying Zang, *D.H. Lawrence's Philosophy of Nature: An Eastern View* (Bloomington, IN: Trafford Publishing, 2011), p. 7.

65   The sabinenes seem: "The Forest and Human Health Issues in Korean Forest Policy and Research," topic paper, Korea Forest Research Institute, Oct. 27, 2014.

66   Flying out of poverty: This is based on the World Bank's most recent ranking, found here: http://databank.worldbank.org/data/download/GDP.pdf, accessed June 2015.

66   98 percent of South Koreans graduate: Tudor, Kindle location 1954.

67   In a country where: Tudor, Kindle location 1939.

67   *sanshin*, the mountain spirit: Hong, Kindle locations 740, 757.

67   Trees, too, have long been: Tudor, Kindle location 498.

67   which means body and soil are one: Hong, Kindle location 726.

73   1 trillion odors: Caroline Bushdid et al., "Humans Can Discriminate More Than 1 Trillion Olfactory Stimuli," *Science,* vol. 343, no. 6177 (2014): pp. 1370–72.

73   The researchers measured: Lilianne R. Mujica-Parodi et al., "Chemosensory Cues to Conspecific Emotional Stress Activate Amygdala in Humans," *PLoS ONE,* vol. 4, no. 7 (2008), published online, e6495.

73   Svante Pääbo is the Swedish: This interview with Pääbo about human smell is available online through Cold Spring Harbor Laboratory's DNA Learning Center website: http://www.dnalc.org/view/15149-Human-smell-receptors-Svante-Paabo.html, accessed Nov. 2014.

74   what about us?: For more on the domestication of humans, see Razib Khan, "Our Cats, Ourselves," *New York Times,* Nov. 24, 2014, accessed Nov. 2014.

75   2.1 million premature deaths annually: Tami C. Bond et al., "Bounding the Role of Black Carbon in the Climate System: A Scientific Assessment," *Journal of Geophysical Research: Atmospheres,* vol. 118, no. 11 (2013): pp. 5380–552.

75   smog-choked Mexico City: Calderón-Garcidueñas et al., "Air Pollution, Cognitive Deficits and Brain Abnormalities: A Pilot Study with Children and Dogs," *Brain and Cognition,* vol. 68, no. 2 (2008): pp. 117–27.

76   Nineteen percent of Americans: Gregory M. Rowangould, "A Census of the U.S. Near-Roadway Population: Public Health and Environmental Justice Considerations," *Transportation Research Part D: Transport and Environment,* vol. 25 (2013): pp. 59–67. The study also mentioned that "greater traffic volume and density are associated with larger shares of non-white residents and lower median household incomes," on a national level. Additionally, counties with residents living near high-volume roads often do not have an air-quality monitor in the same area.

76   rose petals to lure Marc Antony: Diane Ackerman, *A Natural History of the Senses* (New York: Vintage Books, 1995), p. 36.

76   pleasant smells trigger "approach behavior": Paula Fitzgerald Bone and Pam Scholder Ellen, "Scents in the Marketplace: Explaining a Fraction of Olfaction," *Journal of Retailing,* vol. 75, no. 2 (1999): pp. 243–262.

76   If a store smells good: Rob W. Holland, Merel Hendriks, and Henk Aarts, "Smells Like Clean Spirit: Nonconscious Effects of Scent on Cognition and Behavior," *Psychological Science,* vol. 16, no. 9 (2005): pp. 689–93.

76   People assigned to a room: Katie Liljenquist, Chen-Bo Zhong, and Adam D. Galinsky, "The Smell of Virtue: Clean Scents Promote Reciprocity and Charity," *Psychological Science,* vol. 21, no. 3 (2010): pp. 381–83.

77   The so-called "pinosylvin": Mi-Jin Park, "Inhibitory Effect of the Essential Oil from Chamaecyparis obtuse on the Growth of Food-Borne Pathogens," *Journal of Microbiology,* vol. 48, no. 4. (2010): pp. 496–501.

77   Although aromatherapy is the most popular alternative: Yuk-Lan Lee et al., "A Systematic Review of the Anxiolytic Effects of Aromatherapy in People with Anxiety Symptoms," *Journal of Alternative and Complementary Medicine,* vol. 17, no. 2 (2011): p. 106.

77   "a safe and pleasant intervention": Lee, p. 107.

77   significantly less anxiety using "aromasticks": Jacqui Stringer and Graeme Donald, "Aromasticks in Cancer Care: An Innovation Not to Be Sniffed At," *Complementary Therapies in Clinical Practice,* vol. 17, no. 2 (2011): pp. 116–21.

77   Other studies have reported: Toshiko Atsumi and Keiichi Tonosaki, "Smelling Lavender and Rosemary Increases Free Radical Scavenging Activity and Decreases Cortisol Level in Saliva," *Psychiatry Research* ,vol. 150, no. 1 (2007): pp. 89–96, and Yumi Shiina et al., "Relaxation Effects of Lavender Aromatherapy Improve Coronary Flow Velocity Reserve in Healthy Men Evaluated by Transthoracic Doppler Echocardiography." *International Journal of Cardiology,* vol. 129, no. 2 (2008): pp. 193–97.

77   In one survey of 400 Londoners: George MacKerron and Susana Mourato, "Life Satisfaction and Air Quality in London," *Ecological Economics,* vol. 68, no. 5 (2009): pp. 1441–53.

CHAPTER 4: BIRDBRAIN

85   "Most people": From Hemingway's letter of advice to a young writer, reported in Malcolm Cowley, "Mister Papa," *Life,* Jan. 10, 1949, p. 90.

86   "Noise" is unwanted sound: Kurt Fristrup, senior scientist, National Park Service, from a talk at the AAAS conference in San Jose, California, Feb. 16, 2015.

86   Traffic on roads in the United States: Jesse R. Barber et al., "Conserving the Wild Life Therein: Protecting Park Fauna from Anthropogenic Noise," *Park Science,* vol. 26, no. 3 (Winter 2009–10), p. 26.

86 The number of passenger flights: The number of flights, as well as other data, are available going back to 2002 through the Bureau of Transportation's Tran-Stats website, accessible here: http://www.transtats.bts.gov/Data_Elements .aspx?Data=1, accessed June 2015.

86 30,000 commercial aircraft: From the National Oceanic and Atmospheric Administration, http://sos.noaa.gov/Datasets/dataset.php?id=44, accessed 6/16/16.

86 90 percent increase in air traffic: FAA Aerospace Forecast Fiscal Years 2012–2032, quoted in Gregory Karp, "Air Travel to Nearly Double in Next 20 Years, FAA Says," *Chicago Tribune*, March 8, 2012, accessed Feb. 2015.

86 about 30 decibels: Human development has increased noise levels by 30 decibels, from the National Park Service, see graphic at http://media.thenews tribune.com/smedia/2014/05/17/16/18/1nMDoK.HiRe.5.jpg, accessed Feb. 2015.

87 decibel levels between 55 and 60: Average decibels for my neighborhood, the D.C. Palisades, from the 2013 Annual Aircraft Noise Report of the Metropolitan Washington Airports Authority, http://www.mwaa.com/file/2013_noise_ report_final2.pdf, accessed Feb. 2015.

88 It was so airtight: I read about Carlyle's attic in Don Campbell and Alex Doman, *Healing at the Speed of Sound: How What We Hear Transforms Our Brains and Our Lives* (New York: Hudson Street Press, 2011), Kindle location 566.

88 In one study that lasted: Barbara Griefahn et al., "Autonomic Arousals Related to Traffic Noise During Sleep," *Sleep*, vol. 31, no. 4 (2008): p. 569.

89 It's not uncommon in the animal world: Barber, p. 26.

89 That's enough to reduce the distance: Barber, p. 26.

89 it takes them longer to find males: Barber, p. 29.

89 Nerve cells pick up these perturbations: For a good description of how sound travels through the brain, see Daniel Levitin, *This Is Your Brain on Music* (New York: Penguin Group, 2006), pp. 105–6.

89 As to the perennial question: Levitin, p. 29.

90 But there is no thing called sound: For more on Berkeley's question, see Levitin, p. 24.

90 In a study of 2,000 men: These studies of noise and hypertension are described in Martin Kaltenbach, Christian Maschke, and Rainer Klinke. "Health Consequences of Aircraft Noise." *Dtsch Arztebl Int*, vol. 105, no. 31-32 (2008): pp. 548–56.

91 Their systolic blood pressure went up: The Munich airport study: Gary Evans et al., "Chronic Noise Exposure and Physiological Response: A Prospective Study of Children Living Under Environmental Stress," *Psychological Science*, vol. 9, no. 1 (1998): pp. 75–77.

91 As the authors of an important review paper: Kaltenbach et al., 2008.

92 "the world's first anti-noise martyr": Campbell and Doman, *Healing at the Speed of Sound,* Kindle location 2466.

94 Visitors hearing loud vehicle noise: David Weinzimmer et al., "Human Responses to Simulated Noise in National Parks," *Leisure Sciences: An Interdisciplinary Journal,* vol. 36, issue 3 (2014): pp. 251–67.

94 Opposite effects are seen in cities: Subjects in cities rate them as more attractive when listening to birdsong: Marcus Hedblom et al., "Bird Song Diversity Influences Young People's Appreciation of Urban Landscapes," *Urban Forestry & Urban Greening,* vol. 13, no. 3 (2014): pp. 469-474. Another interesting factoid is that hearing other people's voices impairs park visitors' memories. See Jacob A. Benfield et al., "Does Anthropogenic Noise in National Parks Impair Memory?" *Environment and Behavior,* vol. 42, no. 5 (2010): pp. 693–706.

98 John Ruskin wrote: Ruskin quote from "Unto This Last" (1862), cited in Jonathan Bate, *Romantic Ecology: Wordsworth and the Ecological Tradition* (London: Rutledge, 1991), preface.

98 Darwin devoted ten pages to birdsong: On Darwin, I gathered these page counts from Gordon H. Orians, *Snakes, Sunrises, and Shakespeare: How Evolution Shapes Our Loves and Fears* (Chicago: University of Chicago Press, 2014), Kindle location 1877.

98 British Petroleum gas stations recently began playing birdsong: Denise Winterman, 'The Surprising Uses for Birdsong', *BBC Magazine,* May 8, 2013, http://www .bbc.com/news/magazine-22298779, accessed February 2015.

98 In fact, birdsong has some: Factoids on the brown thrasher and others from http://www.pbs.org/lifeofbirds/songs/, accessed February 2015.

99 This is because humans and birds: On the comparison between bird brain structures and the basal ganglia, see Johan J. Bolhuis et al., "Twitter Evolution: Converging Mechanisms in Birdsong and Human Speech," *Nature Reviews Neuroscience,* vol. 11, no. 11 (2010): pp. 747–59.

99 It's well recognized that music triggers emotions: For more on coevolution and the fascinating similarities in gene expression and brain structures between birds and humans, see Bolhuis, but also Cary H. Leung et al., "Neural Distribution of Vasotocin Receptor MRNA in Two Species of Songbird," *Endocrinology,* vol. 152, no. 12 (2011): pp. 4865–81, and Michael Balter, "Animal Communication Helps Reveal Roots of Language," *Science,* vol. 328, no. 5981 (2010): pp. 969–71.

CHAPTER 5: BOX OF RAIN

105 "[When] the myopia": Juler quote from Elie Dolgin, "The Myopia Boom" *Nature,* vol. 519, no. 7543 (2015): pp. 276–78, accessed March 2015.

105   "She promised us south rooms": E. M. Forster, *A Room with a View* (New York: Knopf, 1922), p. 13.

107   Nightingale's famous nursing textbook: Florence Nightingale, *Notes on Nursing: What It Is, and What It Is Not* (New York: D. Appleton & Co., 1860), accessed at http://digital.library.upenn.edu/women/nightingale/nursing/nursing.html in April 2015.

108   One of the first people: "View Through a Window May Influence Recovery," *Science,* vol. 224, no. 4647 (1984): pp. 224–25.

108   prisoners in Michigan whose cells: E. O. Moore, "A Prison Environment's Effect on Health Care Service Demands," *Journal of Environmental Systems,* vol. 11 (1981): pp. 17–34.

109   the brutalist Robert Taylor housing project: For the series of Robert Taylor Homes studies, see Frances E. Kuo, "Coping with Poverty: Impacts of Environment and Attention in the Inner City," *Environment & Behavior,* vol. 33, no. 1 (2001): pp. 5–34; Frances E. Kuo and William C. Sullivan, "Aggression and Violence in the Inner City: Effects of Environment via Mental Fatigue," *Environment & Behavior,* Special Issue, vol. 33 no. 4 (2001): pp. 543–71.

110   Analyzing 98 buildings over two years: Frances E. Kuo and William C. Sullivan, "Environment and Crime in the Inner City: Does Vegetation Reduce Crime?" *Environment & Behavior,* vol. 33, no. 3 (2001): pp. 343–67.

111   The greener-courtyard residents: Frances E. Kuo et al., "Fertile Ground for Community: Inner-City Neighborhood Common Spaces," *American Journal of Community Psychology,* vol. 26, no. 6 (1998): pp. 823–51.

111   For some reason, social psychologists: For the road rage study, see Jean Marie Cackowski, and Jack L. Nasar, "The Restorative Effects of Roadside Vegetation Implications for Automobile Driver Anger and Frustration," *Environment and Behavior,* vol. 35, no. 6 (2003): pp. 736–51.

111   In these studies: The Dutch study is Jolanda Maas et al., "Social Contacts as a Possible Mechanism Behind the Relation Between Green Space and Health," *Health and Place,* vol. 15, no. 2 (2009): pp. 586–95. The office plant study is Netta Weinstein, Andrew K. Przybylski, and Richard M. Ryan, "Can Nature Make Us More Caring? Effects of Immersion in Nature on Intrinsic Aspirations and Generosity," *Personality and Social Psychology Bulletin,* vol. 35, no. 10 (2009): pp. 1315–29.

112   Several years ago Taylor wrote: Richard Taylor, "The Curse of Jackson Pollock: The Truth Behind the World's Greatest Art Scandal," *Oregon Quarterly,* vol. 90, no. 2 (2010), http://materialscience.uoregon.edu/taylor/CurseOfJackson Pollock.pdf, accessed March 2015.

113   Arthur C. Clarke described the Mandelbrot set: The quote is from a documentary presented by Arthur C. Clarke, *The Colours of Infinity,* directed by Nigel

Lesmoir-Gordon (1995), available on YouTube: https://www.youtube.com/watch?v=Lk6QU94xAb8, accessed June 2015.

114   He and Caroline Hagerhäll: Caroline M. Hagerhäll et al., "Fractal Dimension of Landscape Silhouette Outlines as a Predictor of Landscape Preference," *Journal of Environmental Psychology*, vol. 24, no. 2 (2004): pp. 247–55.

114   To find out, they used EEG: For a fuller discussion of the EEG study, see Richard Taylor et al., "Perceptual and Physiological Responses to Jackson Pollock's Fractals," *Frontiers in Human Neuroscience*, vol. 5 (2011): pp. 60–70.

115   Taylor believes our brains recognize that: For more on fractals in art and nature, see Branka Spehar and Richard P. Taylor, "Fractals in Art and Nature: Why Do We Like Them?" *Human Vision and Electronic Imaging XVIII*, March 14, 2013, published online.

115   Pollock's favored dimension is similar: Taylor, p. 60.

116   this D range elicits our best: B. E. Rogowitz and R. F Voss, "Shape Perception and Low Dimension Fractal Boundary Contours," in B. E. Rogowitz and J. Allenbach, eds., *Proceedings of the Conference on Human Vision: Methods, Models and Applications, SPIE/SPSE Symposium on Electron Imaging, 1990*, vol. 1249, pp. 387–94), cited in Hagerhäll (2004).

116   "The stress-reduction is triggered": Quote from Richard Taylor, "Human Physiological Responses to Fractals in Nature and Art: a Physiological Response," author page at http://materialscience.uoregon.edu/taylor/rptlinks2.html, accessed March 2015.

117   Long before fractals, Beethoven: Beethoven wrote the resonance sentences in a letter to Therese Malfatti, his student and love interest, after completing Symphony No. 6 in F Major, titled *Pastoral*, 1808, cited here: http://worldhistoryproject.org/1808/beethoven-finishes-his-sixth-symphony, accessed March 2015.

119   "we will suffer physical and psychological costs": Peter H. Kahn, Rachel L. Severson, and Jolina H. Ruckert. "The Human Relation with Nature and Technological Nature," *Current Directions in Psychological Science*, vol. 18, no. 1 (2009): p. 41.

124   Since red makes us vigilant: We walk down red corridors faster . . . Peter Aspinall, personal communication, June 2014.

124   "If you want to make": Humphrey quote from Natalie Angier, "How Do We See Red? Count the Ways," *New York Times*, Feb. 6, 2007, http://www.nytimes.com/2007/02/06/science/06angi.html, accessed April 2015.

124   But pink, interestingly, has the opposite effect: For more on the psychology of color, see Adam Alter's aptly named *Drunk Tank Pink* (New York: Penguin Group, 2013).

124   Berger writes in *The Sense of Sight*: The John Berger quote comes from Diane

Ackerman's *A Natural History of the Senses* (New York: Random House, 1990), p. 177.

124   In the app, straight and jagged lines: A fuller description of the visual proper-
      ties that trigger restoration can be found in D. Valtchanov and C. Ellard, "Cog-
      nitive and Affective Responses to Natural Scenes: Effects of Low Level Visual
      Properties on Preference, Cognitive Load and Eye-Movements," *Journal of
      Environmental Psychology*, vol. 43 (2015): pp. 184–95.

125   the same region Taylor found stimulated: The other studies implicating the ven-
      tral striatum and parahippocampus using fMRI include Xiaomin Yue et al., "The
      Neural Basis of Scene Preferences," *Neuroreport,* vol. 18, no. 6 (2007): pp. 525–29.

125   craving the "visual opium" of a sunset: Ackerman, p. 255.

125   According to Valtchanov: For more on Valtchanov's visuospatial theory, see
      Deltcho Valtchanov, "Exploring the Restorative Effects of Nature: Testing a
      Proposed Visuospatial Theory," diss., University of Waterloo, 2013.

CHAPTER 6: YOU MAY SQUAT DOWN AND FEEL A PLANT

131   "The faint whisper": Jansson quote from *Moominvalley in November* (New
      York: Macmillan, 2014), p. 26, first published in English in 1945.

138   They get five-week vacations: Rebecca Ray, Milla Sanes, and John Schmitt,
      "No-Vacation Nation Revisited" (Center for Economic and Policy Research,
      2013), p. 5, accessible at http://www.cepr.net/documents/publications/no-
      vacation-update-2013-05.pdf, accessed June 2015;and "Annual Holiday"
      (Ministry of Employment and the Economy, February 11, 2010), accessible
      at https://www.tem.fi/en/work/labour_legislation/annual_holiday, accessed
      June 2015.

138   as well as paid one-year parental leave: Details of Finnish parental leave can
      be found at http://europa.eu/epic/countries/finland/index_en.htm, accessed
      June 2015.

CHAPTER 7: GARDEN OF HEDON

149   "Clearings. That's what I needed": Quote is from Helen Macdonald, *H is for
      Hawk.* (New York: Random House, 2014).

149   In the Gaelic poem "Hallaig": The haunting audio clip of the poem, read in
      Gaelic, can be found here: http://www.edinburghliterarypubtour.co.uk/
      makars/maclean/hallaig.html, accessed April 2015.

149   Weet, williwaw, crizzle: All from Robert McFarlane's *Landmarks* (London:
      Penguin UK, 2015).

150   In some neighborhoods a man: The information on Glasgow life expectancy

comes from the World Health Organization: http://www.who.int/bulletin/volumes/89/10/11-021011/en/, accessed April 2015.

150 The main cause: *Richard J. Finlay, Modern Scotland 1914–2000* (London: Profile Books, 2004).

152 we expended about 1,000 kilocalories: The kilocalorie figures are cited in Jo Barton and Jules Pretty, "What Is the Best Dose of Nature and Green Exercise for Improving Mental Health? A Multi-Study Analysis," *Environmental Science & Technology*, vol. 44, no. 10 (2010): p. 3947.

153 Walking is the most popular sport in Scotland: From "Let's Get Scotland Walking: The National Walking Strategy," government report (2014), http://www.gov.scot/Resource/0045/00452622.pdf, accessed April 2015.

154 In other words, there was something protective: Richard Mitchell and Frank Popham, "Effect of Exposure to Natural Environment on Health Inequalities: An Observational Population Study," *Lancet,* vol. 372 (2008): pp. 1655–60.

155 "40 percent less than those with the worst access": Mitchell quotes on the AJPM study are from his blog: http://cresh.org.uk/2015/04/21/more-reasons-to-think-green-space-may-be-equigenic-a-new-study-of-34-european-nations/, accessed April 2015. The study itself is Richard J. Mitchell et al., "Neighborhood Environments and Socioeconomic Inequalities in Mental Well-Being," *American Journal of Preventive Medicine*, vol. 49, issue 1 (2015): pp. 80–84.

156 the percentage of Scotland covered by woodland: Martin Williams, "Hopes for Forestry Scheme to Branch Out," *The Herald* (Edinburgh), June 4, 2013. http://www.heraldscotland.com/news/home-news/hopes-for-forestry-scheme-to-branch-out.21253639, accessed May 2014.

161 Benjamin Rush, who first popularized the idea: Benjamin Rush quote from *Benjamin Rush, Medical Inquiries and Observations upon Diseases of the Mind* (Philadelphia: Kimber & Richardson, 1812), p. 226, accessed at https://archive.org/stream/medicalinquiries1812rush#page/n7/mode/2up, accessed May 2015.

162 "It was as though": Johan Ottosson, "The Importance of Nature in Coping," diss., Swedish University of Agricultural Sciences, 2007, p. 167.

164 Its motto could be the Emerson quote: Emerson vegetable quote from Ralph Waldo Emerson, *Nature* (Boston: James Munroe & Co., 1836), p. 13. A digital version of the original essay is available here: https://archive.org/details/naturemunroe00emerrich, accessed June 2015.

166 For some other cool UK studies about happiness, health and coastlines, see M.P. White et al., "Coastal Proximity, Health and Well-being: Results from a Longitudinal Panel Survey," *Health Place*, vol. 23 (2013): pp. 97–103; and B.W. Wheeler et al., "Does Living by the Coast Improve Health and Wellbeing?" *Health Place*, vol. 18 (2012): pp. 1198–201.

167 Other good walking studies include Melissa Marselle et al., "Examining Group

Walks in Nature and Multiple Aspects of Well-Being: A Large-Scale Study," *Ecopsychology*, vol. 6, no. 3 (2014): pp. 134–147, and Melissa Marselle et al., "Walking for Well-Being: Are Group Walks in Certain Types of Natural Environments Better for Well-Being than Group Walks in Urban Environments?" *International Journal of Environmental Research and Public Health*, vol. 10, no. 11 (2013): pp. 5603–28.

CHAPTER 8: RAMBLING ON

169   "When we walk": From Henry David Thoreau, "Walking," in *The Writings of Henry David Thoreau*, Riverside ed. (Boston: Houghton Mifflin, 1893), p. 258.

169   Gros writes in *A Philosophy of Walking*: Gros is quoted in Carole Cadwalladr, "Frédéric Gros: Why Going for a Walk Is the Best Way to Free Your Mind," *The Guardian*, April 19, 2014, http://www.theguardian.com/books/2014/apr/20/frederic-gros-walk-nietzsche-kant, accessed May 2015.

170   Anticipating the exercise/nature debate: Henry David Thoreau, "Walking," Kindle location 54.

170   He also wrote, in his essay "Walking": Thoreau, Kindle location 33.

170   "To you, clerk": Velsor Mose (Walt Whitman), "Manly Health and Training, with Off-Hand Hints Toward Their Conditions," ed. Zachary Turpin, *Walt Whitman Quarterly Review* 33 (2016), p. 189.

171   Hartman's own history: Hartman's relocation story is told in Jon Nordheimer, "15 Who Fled Nazis as Boys Hold a Reunion," *New York Times*, July 28, 1983.

172   how it "interfused" with the mind: Wordsworth external mind quotes are from the First Book of *The Recluse*.

172   a "savage torpor": Savage torpor, from the preface to *Lyrical Ballads*, quoted in James A. W. Heffernan, "Wordsworth's London: The Imperial Monster," *Studies in Romanticism*, vol. 37, no. 3 (1998): pp. 421–43.

174   He also believed: For a good overview of Berger's quest and legacy, see David Millett, "Hans Berger: From Psychic Energy to the EEG," *Perspectives in Biology and Medicine*, vol. 44, no. 4 (2001): pp. 522–42.

174   walk around Edinburgh: The Edinburgh EEG study: Peter Aspinall et al., "The Urban Brain: Analysing Outdoor Physical Activity with Mobile EEG," *British Journal of Sports Medicine* (2013), published online, bjsports-2012-091877.

177   forty minutes of moderate walking: For Kramer's exercise studies, see Charles H. Hillman et al., "Be Smart, Exercise Your Heart: Exercise Effects on Brain and Cognition," *Nature Reviews Neuroscience,* vol. 9, no. 1 (2008): pp. 58–65, and Kirk I. Erickson et al., "Exercise Training Increases Size of Hippocampus and Improves Memory," *Proceedings of the National Academy of Sciences*, vol. 108, no. 7 (2011): pp. 3017–22.

177   Kramer was intrigued: The Stanford walking study is Marily Oppezzo and
      Daniel L Schwartz, "Give Your Ideas Some Legs: The Positive Effect of Walking
      on Creative Thinking," *Journal of Experimental Psychology: Learning, Mem-
      ory and Cognition*, vol. 40, no. 4 (2014).

180   The Bratman "dish" study: Greg Bratman et al., "The Benefits of Nature Expe-
      rience: Improved Affect and Cognition," *Landscape and Urban Planning*, vol.
      138 (2015), pp. 41–50.

181   "The results suggest": From Gregory N. Bratman et al., "Nature Experience
      Reduces Rumination and Subgenual Prefrontal Cortex Activation," *Proceed-
      ings of the National Academy of Sciences*, vol. 112, no. 28 (2015): p. 8567.

CHAPTER 9: GET OVER YOURSELF: WILDERNESS, CREATIVITY AND THE POWER OF AWE

Some of the information in this chapter originally appeared in different form in Florence
Williams's *National Geographic* story "This Is Your Brain on Nature," January 2016.
Calvin and Hobbes quote from Bill Watterson, *The Complete Calvin and Hobbes* (River-
side, NJ: Andrews McNeel, vol.3, 2005), p. 370. Bachelard quote, cited in Michael Pol-
lan, *Cooked: A Natural History of Transformation* (New York: Penguin Press, 2013), p.
109. Ellen Meloy quotes from her lovely work of memoir-slash-natural history, *The Last
Cheater's Waltz* (New York: Henry Holt, 1999), pp. 7, 107. Ed Abbey's chapter title from
Desert Solitaire: A Season in the Wilderness (Tucson: University of Arizona Press, 1988).

187   "Look at all the stars!": Bill Watterson, *The Complete Calvin and Hobbes*, Vol. 3
      (Riverside, NJ: Andrews McMeel, 2005), p. 370.

194   "The passion caused": From Edmund Burke, *A Philosophical Enquiry into the
      Origin of our Ideas of the Sublime and Beautiful* (London: University of Notre
      Dame Press, 1968), p. 57.

195   For more on the origins of the word "awe," see Dacher Keltner, *Born to Be Good*
      (New York: W. W. Norton, 2009), p. 257.

195   For more on Burke's influence on Kant and Diderot, see the introduction by
      James T. Boulton in Burke, 1968 ed., p. cxxv ff.

196   "inverse P.T.S.D.": Cited in Michael Pollan, "The Trip Treatment," *New Yorker*,
      Feb. 19, 2015, http://www.newyorker.com/magazine/2015/02/09/trip-treatment,
      accessed Oct. 2, 2015.

196   The Piff and Keltner study: Paul K. Piff et al., "Awe, the Small Self, and Pro-
      social Behavior," *Journal of Personality and Social Psychology*, vol. 108, no. 6
      (2015): p. 883.

197   The cytokine study is Jennifer E. Stellar et al., "Positive Affect and Markers of
      Inflammation: Discrete Positive Emotions Predict Lower Levels of Inflamma-
      tory Cytokines," *Emotion*, vol. 15, no. 2 (2015).

198   For more about Darwin on compassion and the emotion of awe generally, I rec-
       ommend Keltner's *How to Be Good*. A more academic summary can be found
       in Michelle N. Shiota, Dacher Keltner, and Amanda Mossman, "The Nature of
       Awe: Elicitors, Appraisals, and Effects on Self-Concept," *Cognition and Emo-
       tion,* vol. 21, no. 5 (2007): pp. 944–63.

200   Nearly half of all Americans: J. Carroll, "Time Pressures, Stress Common for
       Americans" a Gallup-Time Poll from 2008, cited in Rudd, 2012.

200   For more on awe and time perception, see Melanie Rudd et al., "Awe Expands
       People's Perception of Time, Alters Decision Making, and Enhances Well-
       Being," *Psychological Science* vol. 23, no. 10 (2012). For more on awe and gen-
       erosity, see Netta Weinstein et al., "Can Nature Make Us More Caring? Effects
       of Immersion in Nature on Intrinsic Aspirations and Generosity," *Personality
       and Social Psychology Bulletin,* vol. 35, no. 10 (2009): pp. 1315–40.

CHAPTER 10: WATER ON THE BRAIN

203   "Oh Eeyore, you are wet!": A. A. Milne, *The House at Pooh Corner*, deluxe ed.
       (New York: Dutton, 2009), p. 101.

203   "Between every two": From Muir's marginalia in his copy of *Prose Works by
       Ralph Waldo Emerson,* vol. 1 (this volume resides in the Beinecke Rare Book
       and Manuscript Library of Yale University). Cited in "Quotations from John
       Muir," selected by Harold Wood, http://vault.sierraclub.org/john_muir_
       exhibit/writings/favorite_quotations.aspx, accessed April 12, 2016.

203   "I Sliped & bruised my leg very much": Lewis and Clark account from lew-
       is-clark.org/content/content-article.asp?ArticleID=1790, accessed Sept. 2014.

207   Surgeons in World War I: For a look at the role of plastic surgery in World War I,
       see Sheryl Ubelacker, "Unprecedented Injuries from First World War Spawned
       Medical Advances Still Used Today," *Canadian Press* (via Postmedia's World
       War 1 Centenary site), Sept. 23, 2014, http://ww1.canada.com/battlefront/
       unprecedented-injuries-from-first-world-war-spawned-medical-advances-
       still-used-today, accessed June 2015. For an overview of the effects of mustard
       gas, see "Facts About Sulfur Mustard," Centers for Disease Control, May 2,
       2013, http://www.bt.cdc.gov/agent/sulfurmustard/basics/facts.asp, accessed
       June 2015.

208   Union soldiers after the battle: Olmsted quote from Rybczynski, Kindle edition
       location 3244.

208   PTSD wasn't officially named and recognized: Matthew J. Friedman, "PTSD His-
       tory and Overview," U.S. Department of Veterans Affairs, March 2, 2014, http://
       www.ptsd.va.gov/PTSD/professional/PTSD-overview/ptsd-overview.asp.

208   Among veterans, that figure: "Witness Testimony of Karen H. Seal, M.D.,

MPH," House Committee on Veterans' Affairs, June 14, 2011, http://Veterans. house.gov/prepared-statement/prepared-statement-karen-h-seal-md-mph-department-medicine-and-psychiatry-san, as quoted in David Scheinfleld, "From Battlegrounds to the Backcountry: The Intersection of Masculinity and Outward Bound Programming on Psychosocial Functioning for Male Military Veterans," diss., University of Texas at Austin, 2014, p. 27.

208   They are two to four times: Gail Gamache, Robert Rosenheck, and Richard Tessler, "Overrepresentation of Women Veterans Among Homeless Women," *American Journal of Public Health*, vol. 93, no. 7 (2003): pp. 1132–36.

211   In frightened lab animals: For the role of GCs in memory: J-F. Dominique et al., "Stress and Glucocorticoids Impair Retrieval of Long-Term Spatial Memory," *Nature*, vol. 394 (1998): pp. 787–90. For the hippocampus: Nicole Y.L. Oei et al., "Glucocorticoids Decrease Hippocampal and Prefrontal Activation During Declarative Memory Retrieval in Young Men," *Brain Imaging and Behaviour*, vol. 1 (2007): pp. 31–41. For norepinephrine: J. Douglas Bremner, "Traumatic Stress: Effects on the Brain," *Dialogues in Clinical Neuroscience*, vol. 8, no. 4 (2006): pp. 445.

211   Veterans are twice as likely: Jessie L. Bennett et al., "Addressing Posttraumatic Stress Among Iraq and Afghanistan Veterans and Significant Others: An Intervention Utilizing Sport and Recreation," *Therapeutic Recreation Journal*, vol. 48, no. 1 (2014): p. 74.

211   female veterans commit suicide: Matthew Jakupcak et al., "Hopelessness and Suicidal Ideation in Iraq and Afghanistan War Veterans Reporting Subthreshold and Threshold Posttraumatic Stress Disorder," *Journal of Nervous and Mental Disease,* vol. 199, no. 4 (2011): pp. 272–75.

CHAPTER 11: PLEASE PASS THE HACKSAW

Some of the material from this chapter appeared in Florence Williams, "ADHD: Fuel for Adventure," *Outside*, Jan./Feb. 2016, published online Jan. 20, 2016, http://www.outsideonline.com/2048391/adhd-fuel-adventure?utm_source=twitter&utm_medium=social&utm_campaign=tweet, accessed Feb. 22, 2016.

221   "Childhood is, or has been": From "Manhood for Amateurs: The Wilderness of Childhood," *New York Review of Books*," July 19, 2009, www.nybooks.com/articles/archives/2009/jul/16/manhood-for-amateurs-the-wilderness-of-childhood/, accessed July 17, 2015.

224   A recent advertisement for an ADHD drug: Mentioned in Richard Louv's blog post, "NATURE WAS MY RITALIN: What the New York Times Isn't Telling You About ADHD: The New Nature Movement," http://blog.childrenandnature.org/2013/12/16/nature-was-my-ritalin-what-the-new-york-times-isnt-telling-you-about-adhd/, accessed July 20, 2015.

225 Olmsted hated school: From Witold Rybczynski, *A Clearing in the Distance: Frederick Law Olmsted and America in the 19th Century* (New York: Scribner, 1999), Kindle edition location 417. Quote to principal from Kindle edition, location 296.

226 Kuo ADHD studies: see A. Faber Taylor et al., "Coping with ADD: The Surprising Connection to Green Play Settings," *Environment and Behaviour*, vol. 33 (Jan. 2001): pp. 54–77.

226 ADHD kids playing in a park study: Andrea Faber Taylor and Frances E. Ming Kuo, "Could Exposure to Everyday Green Spaces Help Treat ADHD? Evidence from Children's Play Settings," *Applied Psychology: Health and Well-Being*, vol. 3, no. 3 (2011): pp. 281–303.

226 The Barcelona study: Elmira Amoly et al., "Green and Blue Spaces and Behavioral Development in Barcelona Schoolchildren: The Breathe Project," *Environmental Health Perspectives* (Dec. 2014), pp. 1351–58.

227 Kuo and Taylor's 2004 study: Frances E. Kuo and Andrea Faber Taylor, "A Potential Natural Treatment for Attention-Deficit/Hyperactivity Disorder: Evidence from a National Study," *American Journal of Public Health*, vol. 94, no. 9 (2004).

228 On play and ADHD, see Jaak Panksepp, "Can PLAY Diminish ADHD and Facilitate the Construction of the Social Brain?" *Journal of the Canadian Academy of Child and Adolescent Psychiatry—Journal de l'Académie canadienne de psychiatrie de l'enfant et de l'adolescent*, vol. 16, no. 2 (2007): p. 62.

229 "Children cannot bounce off the walls": Quote by Erin Kenny, cited in David Sobel, "You Can't Bounce Off the Walls if There Are No Walls: Outdoor Schools Make Kids Happier—and Smarter," *YES! Magazine*, March 28, 2014. http://www.yesmagazine.org/issues/education-uprising/the-original-kindergarten?utm_source=FB&utm_medium=Social&utm_campaign=20140328, accessed July 17, 2015.

229 "Everything is good": The Rousseau quote is from *Émile*, cited in Norman Brosterman, *Inventing Kindergarten* (New York: Harry N. Abrams, 1997), p. 19.

230 For more on the tremendous and largely unsung influence of Friedrich Fröbel, see Brosterman, who makes a fascinating case for Fröbelian kindergarten literally catalyzing modern art. Braque, Kandinsky, Le Corbusier and Frank Lloyd Wright all spent years holding cubes and making abstract geometric patterns with Fröbel's materials, and Wright and Le Corbusier in particular directly credit this for their design sense. Brosterman suggests these influences were largely ignored by art historians because they stemmed from the domain of young children and their women teachers.

232 Finns and ADHD: S. L. Smalley et al., "Prevalence and Psychiatric Comorbidity of Attention-Deficit/Hyperactivity Disorder in an Adolescent Finnish Population," *Journal of the American Academy of Child and Adolescent Psychiatry*,

vol. 46, no. 12 (Dec. 2007): pp. 1575–83, cited in Daniel Goleman, "Exercising the Mind to Treat Attention Deficits," *New York Times,* May 12, 2014.

232   A large meta-analysis of dozens: B. A. Sibley et al., "The Relationship Between Physical Activity and Cognition in Children: A Meta-analysis," *Pediatric Exercise Science,* vol. 15, no. 3 (2003): pp. 243–56.

232   The Penn State study on social skills: Damon E. Jones et al., "Early Social-Emotional Functioning and Public Health: The Relationship Between Kindergarten Social Competence and Future Wellness," *American Journal of Public Health,* vol. 105, no. 11 (2015): pp. 2283–90.

233   The 2015 *Pediatrics* study on physical activity in preschoolers: Pooja S. Tandon et al., "Active Play Opportunities at Child Care," *Pediatrics,* May 18, 2015, published online.

233   30 percent of third-graders: Romina M. Barros, et al., "School Recess and Group Classroom Behavior," *Pediatrics,* vol. 123, no. 2 (2009): pp. 431–36.

233   "Containerized kids": See http://www.usatoday.com/news/health/2004-11-05-active_x.htm, accessed Feb. 2, 2016.

233   In 2004, 70 percent of mothers: R. Clements, "An Investigation of the Status of Outdoor Play," *Contemporary Issues in Early Childhood,* vol. 5 (2004): pp. 68–80. Also see S. Gaster, "Urban Children's Access to Their Neighbourhoods: Changes Over Three Generations" (1991), quoted in R. Louv, *Last Child in the Woods* (Chapel Hill, NC: Algonquin Books, 2005), p. 123. On children and exercise, see M. Hillman, J. Adams, and Whitelegg, "One False Move: A Study of Children's Independent Mobility," London: Policy Studies Institute, 1990. And http://www.dailymail.co.uk/news/article-462091/How-children-lost-right-roam-generations.html. On preschool diagnoses of ADHD, see http://www.nytimes.com/2014/05/17/us/among-experts-scrutiny-of-attention-disorder-diagnoses-in-2-and-3-year-olds.html?_r=0, accessed July 18, 2015.

234   Teenagers today have: J. M Twenge et al., "Birth Cohort Increases in Psychopathology Among Young Americans, 1938–2007: A Cross-Temporal Meta-Analysis of the MMPI," *Clinical Psychology Review,* vol. 30 (2010): pp. 145–54, cited in M. Brussoni et al., "Risky Play and Children's Safety: Balancing Priorities for Optimal Child Development," *International Journal of Environmental Research and Public Health,* vol. 9 (2012): pp. 3136–48.

CHAPTER 12: NATURE FOR THE REST OF US

241   "If man is not": Olmsted epigraph quoted in Rybczynski, Kindle location 2776.

241   For more on the idea of Metro sapiens, see Jason Vargo, "Metro Sapiens, an Urban Species," *Journal of Environmental Studies and Sciences,* vol. 4, no. 3 (2014).

241 By 2030, there will be: See R. Dhamodaran, "The Great Migration—India by 2030 and Beyond: Harnessing Technology for Better Urban Transportation in India," a presentation to the Wilson Center, http://www.wilsoncenter.org/sites/default/files/RAMAKRISHNAN%2C%20DHAMODARAN_Presentation.pdf, accessed July 31, 2015.

242 "be anything but a hell": Glaeser quote from http://www.cityjournal.org/2014/24_3_urbanization.html, accessed July 31, 2015.

242 Leyhausen's cat studies and the rat results: Cited in E. O. Wilson, *Sociobiology* (Cambridge, Mass: Harvard University Press, 2000), p. 255.

242 For more about the increased risk of mental disorders in city dwellers, see Florian Lederbogen et al., "City Living and Urban Upbringing Affect Neural Social Stress Processing in Humans," *Nature,* vol. 474, no. 7352 (2011): pp. 498–501.

242 Meanwhile, a study from Portugal: S. Marques and M. L. Lima: "Living in Grey Areas: Industrial Activity and Psychological Health," *Journal of Environmental Psychology*, vol. 31 (2011): 314–22, cited in "The Natural Environments Initiative: Illustrative Review and Workshop Statement," Report, Harvard School of Public Health, Center for Health and the Global Environment, 2014, p. 11.

242 We could use some more resilience: World Health Organization fact sheet, http://www.who.int/mediacentre/factsheets/fs369/en/, accessed Aug. 3, 2015.

243 Singapore is the third-densest: World Bank stats found at http://www.infoplease.com/ipa/A0934666.html, accessed Aug. 1, 2015.

244 On Singapore, see Lee Kuan Yew, *From Third World to First: The Singapore Story: 1965–2000* (Singapore: Times Editions: Singapore Press Holdings, 2000), p. 199.

247 Portland hospital infection study: S. W. Kembel et al., "Architectural Design Influences the Diversity and Structure of the Built Environment Microbiome," *ISME Journal*, vol. 6, no. 8 (Jan. 26, 2012): pp. 648–50.

250 The Donovan ash tree study: Geoffrey H. Donovan et al., "The Relationship Between Trees and Human Health: Evidence from the Spread of the Emerald Ash Borer," *American Journal of Preventive Medicine,* vol. 44, no. 2 (2013): pp. 139–45.

251 For the Toronto study, see Omid Kardan et al., "Neighborhood Greenspace and Health in a Large Urban Center," *Scientific Reports*, vol. 5 (2015): pp. 1–14.

EPILOGUE

253 "But are not exercise": Walt Whitman writing as Mose Velsor, "Manly Health and Training, with Off-Hand Hints Toward Their Conditions," ed. Zachary Turpin, *Walt Whitman Quarterly Review* 33 (2016): p. 212.

255 In the 1870s, he actually: Charles E. Beveridge and Paul Rocheleau, *Frederick Law Olmsted: Designing the American Landscape* (New York: Rizzoli, 1998), p. 45, cited in Carol J. Nicholson, "Elegance and Grass Roots: The Neglected Philosophy of Frederick Law Olmsted," *Transactions of the Charles S. Peirce Society*, vol. XL, no. 2 (Spring 2004), http://www.dathil.com/cadwalader/olmsted _philosophy100.html, accessed Aug. 3, 2015.

ILLUSTRATION CREDITS

Introduction: "Walking the Park," by Jacob DeBailey.

Chapter 1: Lucas Foglia.

Chapter 2: From the Mars Desert Research Station, The Mars Society.

Chapter 3: Lucas Foglia/originally appeared in *National Geographic*, January, 2016.

Chapter 4: Lucas Foglia.

Chapter 5: Lucas Foglia.

Chapter 6: From the Eyes as Big as Plates series by Riita Ikonen and Karoline Hjorth.

Chapter 7: Lucas Foglia.

Chapter 8: McAteer Photography for the Crawick Artland Trust.

Chapter 9: Lucas Foglia.

Chapter 10: Frederick Samuel Dellenbaugh seated in the heart of Lodore Canyon on the Colorado River, photographed by John Hillers while part of the second Powell Survey, 1872, U.S. National Archives and Records Administration record 8464436, War Department. Office of the Chief of Engineers.

Chapter 11: Lucas Foglia.

Chapter 12: Lucas Foglia.